TAVISTOCK INSTITUTE

SOCIAL ENGINEERING THE MASSES

DANIEL ESTULIN

Published by:
Trine Day LLC
PO Box 577
Walterville, OR 97489
1-800-556-2012
www.TrineDay.com
publisher@trineday.net

Library of Congress Control Number: 2015948426

Estulin, Daniel.
Tavistock Institue: Social Engineering for the Masses—1st ed.
p. cm.
Epub (ISBN-13) 978-1-63424-044-4
Kindle (ISBN-13) 978-1-63424-045-1
Print (ISBN-13) 978-1-63424-043-7
1. Tavistock Institute of Human Relations. 2. Conspiracies -- United
States. 3. Brainwashing -- United States. 4. Social psychology. I. Title

FIRST EDITION
10 9 8 7 6 5 4 3 2

Printed in the USA
Distribution to the Trade by:
Independent Publishers Group (IPG)
814 North Franklin Street
Chicago, Illinois 60610
312.337.0747
www.ipgbook.com

Murió la Verdad

Los desastres de la guerra, plate No. 79, (1st edition, Madrid: Real Academia de Bellas Artes de San Fernando, 1863)

CONTENTS

Introduction

Tavistock in Sussex, England, "is the world's center for mass brainwashing and social engineering activities."[1] From a somewhat crude beginning at Wellington House, grew a sophisticated organization that was to shape the destiny of the entire planet, and in the process, change the paradigm of modern society.

In this revolutionary work, which undoubtedly will have the effect of a 500-kiloton nuclear explosion, we uncover both the Tavistock network and the methods of brainwashing and psychological warfare being adapted then and now for application to large-scale social engineering projects.

This is the "Aquarian Conspiracy" as the brainwashers refer to themselves, referring to a super secret 1974 Stanford Research Institute study entitled *Changing Images of Man*. You can think of this book as essentially an anti-brainwashing combat manual. Brainwashing depends on the ignorance of the victims. It is all around us. We can all perceive disintegration of our nations in terms of day to day, personal experiences. However, this is not a coincidence. It is not an accident. What we are witnessing all around us is a planned disintegration of the world economy by the most powerful people in the world. This book on the Tavistock Institute attempts to show that the conspiracy is real, who is behind it, what are its final long term objectives and how we the people can stop them from taking us to Hell.

Beyond utter rage and indignation, the conclusion you, the reader will reach is that the moral, material, cultural and intellectual decay that we are witnessing helplessly every day across the globe, is not accidental, not an act of God who is punishing us for our Earthly misdeeds, but a deliberately induced social crisis.

I repeat, this is not a test. This is real and what's at stake is the future of our planet. Welcome! Make yourselves at home! All the kooks of the world are coming out in the most impressive Witch's Sabbath yet!

1

In the world of smoke and mirrors, there are no flukes, no coincidences and no accidents. This, we prove beyond any reasonable doubt, just as we would have to, if we were disputing this case in a court of law. Only the highest standard of excellence and proof will serve our objectives. The stakes are too high and odds against us are too great. What's in the balance, is the future of the planet, the immortality of the human race and the survival of our species. We will prevail. We will succeed, no matter what the costs. There are no second chances, third options or fourth ways. This is it, and with this work that shall stand the test of time, I have drawn the line in the sand. They will not pass! We will not surrender.

During World War II, "Tavistock was the headquarters of the British Army's Psychological Warfare Bureau, which, through the arrangements of the Special Operations Executive also dictated policy to the United States Armed Forces in the matters of psychological warfare."[2]

Look around you! As the result of a frontal assault on our future by the world's leading social scientists and behaviour engineers, the moorings of national moral purpose collapsed. We the people have surrendered to a morally indifferent irrationality. Make no mistake, everything from the New Left to Watergate to Vietnam to Pentagon Papers to the insane hippies, the anti-war movement and the drug-rock counterculture were also pre-planned social engineering projects.

I repeat, what's under assault are not only our individual human rights, but rather the very institution of the "nation state" republic from the oligarchy's "massive social engineering program conducted through Tavistock Institute for Human Relations and other much larger, integrated network of centers of applied social psychology and social engineering that emerged in the aftermath of World War II."[3] These groups regard us and the principles of nation states as their axiomatic philosophical enemy.

This interlocked juggernaut of evil consists of some of the world's most prestigious centers of knowledge and research such as Stanford Research Center at Stanford University, Rand Corporation, MIT/Sloane, "the Advanced Center of Behavioural Sciences at Palo Alto, the Institute of Social Research at the University of Michigan, the Wharton School of Business at the University of Pennsylvania, the Harvard Business School,"[4]

London School of Economics, the National Training Laboratories, the Hudson Institute, Esalen Institute, the National Institute of Mental Health, the National Institute of Drug Abuse, the Office of Naval Research. There are others, such as the Geneva-based International Foundation for Development Alternatives and Executive Conference Center, the first full-time Age of Aquarius graduate school, charged with teaching behaviour modification for high-level executives from *Fortune*'s top 500 companies. Human zombies placed at top-level management positions to lead us into the New Dark Age of transcendental consciousness. Two foci: first, changes needed in the United States; and second, the global order.

Over the period of half a century, tens of billions of dollars have been allocated by the government of the United States with surreptitious help from think-tanks and foundations aligned with Tavistock to fund the work of these groups.

Every aspect of the mental and psychological life of people on the planet was profiled, recorded and stored into computer systems. Above the closely co-operating groups of social scientists, psychologists, psychiatrists, anthropologists, think-tanks and foundations presides the elite of powerful members of the oligarchy, comprised mostly of the old Venetian Black Nobility.

What is the purpose of these behaviour modifications, you may ask? It is to bring about forced changes to our way of life, without our agreement and without ever realizing what is happening to us. The ultimate goal being "the complete extirpation of mankind's inner sense of identity, the tearing out of mankind's innermost soul, and the placement, in the vacant space, of an artificial, synthetic pseudo-soul."[5] However, in order to change mankind's behaviour away from industrial production into spiritualism and to bring us wilfully into the world of post-industrial era zero growth and zero progress, one must force first a change in mankind's 'self-image,' its fundamental conception of what we are. Thus, the image of man appropriate to that new era must be sought, synthesized and then wired into mankind's brain.

As Professor John McMurtry wrote in his 2001 opening address at the Science for Peace Forum held at the University of Toronto, "totality of rule is not the only parameter of totalitarianism. Limitlessness of power also proceeds from an omnipresent center"[6]. In the new totalitarian movement, "this omnipresent directive force"[7] communicates through behaviour modification and identity change – the dominant nodes of the interlocked system.

Psychological terror is not the essence, "but the punctuation mark of the new totalitarianism's meaning. The money-and-consumption command channel is the secret of the movement's success because it avoids responsibility for its failures. Wall Street prescribed market failures to provide for societies are, instead, always attributed to transcendental forces of "the invisible hand" punishing these societies for alleged sins against "market laws." Thus as catastrophes increasingly befall the majority of the world, the victims are blamed for their new deprivation, misery and oppression. This is a far more effective mode of rule than jackboot terror, which is more overt, but it exposes the system to another form of resistance. To keep the majority in a continual state of inner anxiety works because people are made too busy securing or competing for their own survival to co-operate in mounting an effective response. This too, has Tavistock's signature all over it.

"In the past decade, the entire population of the globe has been kept permanently off-balance with one financial meltdown and transnational trade fiat after another emptying national coffers and overriding rights of domestic self-determination. Populations have been so overwhelmed by the moving juggernaut of economic and environmental crises that a rule of universal insecurity has rendered social majorities paralysed by a low-intensity terror – the necessary condition for any totalitarian movement to continue its advance, because keeping its subjects perpetually off balance is its modus operandi.

"In previous periods of history we have seen draconian measures put through at the national level but we have never seen such an assault on peoples' rights and democratic standards. Each new measure, viewed on its own, may seem an aberration, but a whole host of changes as part on an ongoing continuum constitutes a shift"[8] – towards total enslavement. Total power is a supra terrestrial correlative of global power, which can conceive of no limit to itself.

There are many challenges for us to overcome. When presented with all the facts, when the evidence is neatly laid out on the table, when the conspirators have been unmasked and defrocked and their deeds shown for all the world to see, still, the ordinary citizen screams conspiracy, refusing to believe that such a monstrous, mind-boggling collusion is real... Until, it is too late. Be warned!

We are at the crossroads. And the roads we take will now terminate whether we will live in the XXI century as nation state republics or a subjugated, culled and dehumanized crop of slaves.

Dear reader, the situation is extremely grave. We are fighting a combined effort of some of the most brilliant people in history who scheme against us for the purposes of possessing us. But the human will is immortal. Tyrants have killed millions and yet people fought and eventually won their freedom. Freedom stirs the human heart and fear stills it. Amidst the deafening cacophony of patriotic silence, insurgent voices command attention. Immortality has its moral basis in truth and incorruptibility. It deserves to be given all the support that it can get. It deserves to be fought and died for.

Finally, history teaches by analogy, not identity. The historical experience is not one of staying in the present and looking back. Rather it is one of going back into the past and returning to the present with a wider and more intense consciousness of the restrictions of our former outlook.

Francisco Goya's Plate 79 of Disasters shows the fair maid of Liberty flat on her back, bosom exposed. Ghostly figures play about the corpse while monks dig her grave. Truth has died. *Murió la verdad.* How is that for an alternative? Forewarned is forearmed. It is not up to God to save us, it is up to us. We will never find the right answers if we can't ask the proper questions.

Daniel Estulin
Madrid, May 26, 2011

Endnotes

1. http://www.larouchepub.com/eiw/public/1980/eirv07n18-19800513/eirv07n18-19800513_022-tavistock_mother_of_planned_madn.pdf

2. Ibid

3. Ibid

4. Ibid

5. Ibid

6. "Why is there a war in Afghanistan," John McMurtry, Opening Address, Science for Peace Forum and Teach-In, University of Toronto, December 9, 2001

7. Ibid

8. Ibid

CHAPTER ONE

Counterinsurgency

Techniques for psychological manipulation of society are about as old as humanity itself. Feudal lords, to preserve and consolidate their power, forever have used punishment and torture as dissuasive agents of change. "Even thousands of years ago, it was not the techniques per se, but their conscious application as divide-and-conquer tools, which aided the ruling classes. No matter how anti-human a particular technique or therapeutic approach may be, it is not in itself counterinsurgency. Counterinsurgency cannot proceed merely on horrors; it requires conscious and systematic application by the ruling class, or its dupes."[1] This was achieved by fusing psychology and psychiatry in the 1930s.

"The first massive application of psychology as a conscious weapon took place, in Nazi Germany, especially in reference to eugenics, which played upon the most backward Aryan illusions held by and imposed on portions of the masses. While the cause and development of Nazi butchery stemmed wholly from the world economic collapse, its specific form, eugenics, was devised by the Nazi's favorite theoreticians and technicians – the psychiatrists."[2]

Since then, the "science of the mind," has been converted into the craft of mind destruction. Legitimate, therapeutic approaches have given way to behaviour-modifying pseudo-science in the name of aversion therapy.

This conversion of the mind science was truly shaped by war – "the war of mental genocide waged by the bourgeoisie against the working class."[3] The essential premise of the work of Tavistock is the premise that certain kinds of *democratic* "institutions represent far more efficient instrumentalities for fascist dictatorship than the traditional, straightforwardly"[4] *authoritarian* models. "From the great oil hoax and CIA-style brainwashing, the psychological sciences have followed the route initially outlined in 1945 by Dr. John Rawlings Rees, grand master of psywar counterinsurgency, in his book, *The Shaping of Psychiatry by War*."[5] Rees called for the development of psychiatric *shock troops* in order to develop "methods of political control based upon driving the majority of the human population toward psychosis"[6] through procedures of so-called programmed behavioural modification. He proposed this to render the population submissive under the post-World War II economic world order.

Rees told a group of U.S. Army psychiatrists in 1945, "If we propose to come out into the open and to attack the social and national problems of our day, then we must have shock troops and these cannot be provided by psychiatry based wholly in institutions. We must have mobile teams of psychiatrists who are free to move around and make contacts with the local situation in their particular area."[7]

Rees' logic is clear. For true mental health, there must be a complete transformation of society along the lines of rational selection. But, as he laments in his book, "many don´t see reality this way, including most workers who believe that any method of selection is a mechanism by which the wicked capitalist aims to get more work out of the workers, and that argument dies hard."[8] In the Reesian world-view, such nay-sayers, along with anyone who engages in "strikes" or "subversive activity," are themselves neurotic, desperately "in need of treatment, but unfortunately unable to see that they are ill. In such a world of unwitting neurotics, psychiatry, the only arbiter of sanity, can be exercised only by a cabal in every country, groups of psychiatrists, linked to each other" prepared to muster all their weapons and influence for a move "into the political and governmental field."[9]

Only a "conspiracy of psychiatrists" – as Rees meant when he spoke of his mission – "could build a society where it is possible for people of every social group to have treatment when they need it, even when they do not wish it, without the necessity to invoke the law."[10] For Rees, the construction of that cabal became his lifelong 'mission.' As L. Marcus writes in his excellent investigative piece, "Reesian methods rely, completely and consciously, on the destruction of the mental life of world society and a forced march into universal sadism."[11] Within this lies their affinity – men as beasts whose minds, according to Tavistock, are something to be manipulated and destroyed.

Since then, different methodologies of psychological warfare developed at Tavistock Institute have been the central features of the activities of a world-wide set of interlocking think-tanks in consultative capacity and special commissions, government agencies and corporations, their developmental studies and pilot projects with the clear-cut objective of shaping political techniques of social control. Rees and Tavistock organized their cabal according to public dictum: we are not large but well placed.[12] Rees had a clear understanding of power structures, of organizing key individuals who will in turn spread ideas and influence.

When we speak of psychological warfare we are often speaking of ways to make the enemy afraid, and in order to do this we must understand an enemy's psyche: what makes them love, hate, fight, run. That enemy might be foreign or domestic, an army of men or an enraged mass of workers. And in order to find an effective antidote, Tavistock and company need to understand how this enemy will react under stress: will he fight harder or simply surrender? Or will he start making errors in judgement, winning the war for the enemy, in a manner of speaking? The costliest mistakes of psychological warfare operations are always those made in ignorance of an enemy's mindset. This implies a deep knowledge of human psychology by Reesian "shock troops," a knowledge which is itself a kind of black art. And since this is a war of perceptions, of "world views," it is important that the psychologists and the psychiatrists and the sociologists and the anthropologists, these unidentified little grey men in flannel suits working for Tavistock, understand the impact of art, music, literature and other cultural modes of expression and how world-views are represented by them...

As Peter Levenda writes in *Sinister Forces*, "eventually, the temptation will arise to test some of these principles on the domestic population. After all, with whose mindset are we the most familiar but our own? What better place to test new theories of psychological warfare than among our native population?"[13] As Rees said in 1945, "Wars are not won by killing one's opponents but by undermining or destroying his morale whilst maintaining one's own."[14]

One of the key individuals involved in behaviour modification was psychologist, Kurt Lewin. Lewin was the father of group dynamics and one of Rees's first cadre of recruits who began his career at Cornell University, "where he worked on a systematic series of studies of the effect of social pressure on the eating habits of children."[15] He came to the United States in 1933. A refugee from Nazi Germany, Lewin, like many other German intellectuals was forced out of Germany "not because of any basic political differences but as a sacrifice to Hitler's divide-and-conquer anti-Semitism. Lewin, in fact, is noted for his refinement of the Nazi-formulated leaderless group technique into a sophisticated tool of counterinsurgency."[16] One of the lesser known facets of Lewin's work is related to psychological warfare programs, especially showing proper relations between psychological warfare, target-setting, field operations and evaluative reconnaissance. His first overt assignment was to utilize "group decision-making" in changing food preferences away from "meat" towards "whole-wheat bread" as substitutes.

The following passage from his book *Time Perspective and Morale*, illustrates his understanding of psychological warfare: "One of the main techniques for breaking morale through a 'strategy of terror' consists in exactly this tactic – keep the person hazy as to where he stands and just what he may expect. If in addition frequent vacillations between severe disciplinary measures and promises of good treatment together with spreading of contradictory news, make the 'cognitive structure' of this situation utterly unclear, then the individual may cease to even know when a particular plan would lead toward or away from his goal. Under these conditions even those who have definite goals and are ready to take risks, will be paralyzed by severe inner conflicts in regard to what to do."[17]

Lewin's most significant proposal made during the period of World War II and its immediate aftermath, was his conception of 'fascism with a democratic face.' The common psychopathological feature of all fascist's demand is infantilism who defines himself by his attempts to "impose the principle of the autonomous extended family, and to block out the reality"[18] of the outer world. For example, "nationalism" (mother country), "racialism" (mother), "language group" (mother tongue), "cultural affinity group" (family traditions), "community" (extended family, neighbourhood).[19]

Lewin was the first to realize through close observation of his tested cadres that the imposition of fascist-like forms of small group organization and corporatist "structural reforms" could induce fascist ideology in a subject population.

In a sane and moral society, Lewin's proposals would be used for toilet paper and Lewin himself would have been put under protective, psychiatric care. Instead, he was given a lot of money, U.S. citizenship and a grant from the Rockefellers to craft social engineering projects.

Lewin proposed that through a use of "small group" self-brainwashing techniques, a more efficient form of fascist dictatorship could be established. "The ratio and visibility of a horde of jackbooted enforcers characteristic of the Nazi regime, could be reduced by creating fascist forms of small self-administering 'community groups'. They see themselves as existing by means of their ability as individuals to influence the behaviour of those immediately around them."[20] The result, thought Lewin, would be a more efficient form of fascist regime with the superficial appearance of special democratic forms. In other words, "if the atomised individual's world is converted into a controlled environment which conforms to such 'fascist structural reforms', the victim's mind will discover that only its potential paranoid self provides it with the means for agreement with that controversial environment."[21] In other words, fascism is the world desired in the paranoid dreams of the "Id."

What is undeniable, is that Rees and Tavistock were seriously organizing "a cabal to take over the councils of those who are attempting to re-establish the world after the war."[22] Given the training of military, psychiatric and other hard-core fascist cadres, the establishment of a fascist political order would proceed, according to Rees-Lewin Tavistockian model in the following steps:

1. Break down the existing democratic-constitutional institutions. The military and police forces would be reorganized for "civil action," as they are now in the United States. One of the lesser known actions being readied by the government revolves around the "replacement of ordinary state and local police forces by a national counterinsurgency police force modelled on Hitler's S.D. Gestapo, and such as the Royal Canadian Mounted Police is in Canada."[23] At the same time, existing mass institutions would be destroyed by 'spontaneously' organized insurgency. 'Local

community control' groups would be used to destroy broad-based political institutions. Among these recruits to fascist community control, gangs and counter-gangs of terrorists would propagate crime and mutual terroristic confrontations, with both sides under the control and direction of behind-the-scene intelligence operatives. This programmed insurgency of gangs and counter-gangs, mixed with doses of police-controlled terrorist gangs, create the political conditions in which the majority of the population more readily tolerates or even demands various degrees of military and police government, thus creating your "democratic" fascist regime.

2. Eliminate through subversion, assassination, military intervention, embargoes or popular and "spontaneous" uprisings the regime that has outlived its usefulness, and appoint "democratic" civilian government. The appointed "democratic" government can now function only within the limits defined for it by representatives of the supranational agencies.

The specific topics towards the establishing of fascism with a democratic face are as follows:

1. **Area Population Psychological Studies.** During World War II, the Anglo-American psychological warfare services developed a number of studies of specific neurotic susceptibilities of various national cultures. The most famous of these was the so-called Strategic Bombing Survey. It was conceived as a basis for co-ordinating allied bombing of Germany with propaganda and other psychological warfare campaigns against the morale of various enumerated strata of the Third Reich's population,"[24] and was the precursor of CIA-led Vietnam's 'Operation Phoenix', which was a genocide operation in South Vietnam against supporters of Vietcong. In a nutshell, Strategic Bombing Survey mapped out best ways to destroy the morale of the civilian population at the lowest cost.

2. **The media.** The use of control of major media news and cultural media as instruments of inducing desired forms of partial insanity among large populations. In general, by controlling the editorial policies, the news slant concerning national and international issues, the key news agencies and

main mass circulating media determine what the population generally knows and considers credible. Deliberate and habitual falsification of the information is a way of creating "desensitization effects in the mass population by causing the socially-accredited interpretation of cause-effect relations to violate sensuous rational interpretation of experience. This is furthermore is supplemented by the introduction of programmed subliminal psychological material, whose predetermined effect is to accentuate infantile impulses among targeted portions of the population, such as 'human interest' stories which are relatively more gratifying to the infantile impulses and which de-emphasise a rational and scientific overview."[25]

3. **Local Community Control.** "The object of 'local community control' as a fascist counterinsurgency tactic is to fragment the subject population into relatively hermetic political groupings,"[26] narrowing the scale of the groups by separations according to race, sex, language background, regional background, national origin, recreational interests, age groups and neighbourhood. "Setting such groups into competition against each other under the conditions of general austerity is an effective Lewinite technique for inducing self-brainwashing among these groups and progressive psychological deterioration toward polymorphous perverse pseudo-families and outright clinical psychosis."[27]

"The first degree of brainwashing is accomplished by setting 'local community autonomy' into principled opposition to 'big business,' technology, and progressive programs"[28] meant to improve the lives of people within the community. "Programs with emphasis on technological advancement are denounced as efforts of 'outsider elitist groups' to interfere in the autonomous affairs of the local group. At that point, the 'community group' has become functionally semi-psychotic and clinically paranoid as a group. To the extent that the member restricts their social identity to within such a group, the effort to adjust to the group ideals induces corresponding pathological states in the member.

"By setting such groups into competition, and by splitting the group internally through sex, race, income, etc., the paranoia is intensified, the movement toward semi-psychotic is increased"[29] as smaller and smaller sub groups within the community find themselves in cut-throat hostility toward one another.

4. **The application of task-oriented small group brainwashing techniques to leaderless groups.** "These groups operate on the basis of an environment"[30] of reduction in real-income levels and working conditions. Under the conditions of austerity, the brainwashing consists in getting the workers "to make up for part of the lost standard income-rates by ingeniously speeding themselves up."[31] By recycling the employed and the unemployed and large-scale relocation programs and introduction of "group work incentives" and performance-reward competition among competing groups transforms the small production team into a potentially self-brainwashing group. "Under these conditions, semi-psychosis and psychosis cause the group to 'voluntarily' attain degrees of intensification of labor, which cannot be forced from sane labor. The members of such self-brainwashing leaderless group work teams emulate the 'racehorse' syndrome, driving hysterically toward literally suicidal work-paced. Tavistock Institute and University of Pennsylvania are two of the best-known centers where such experimental practices."[32]

One of the key areas of population control is counterinsurgency. Those familiar with the adage that nobody ever leaves the CIA except six feet under the ground may have wondered how such a control could come to be exerted. The answer lies in part due to the insidious methods of Dr. John Rawlings Rees and his Nazi predecessors. Peter Cuskie explains: "Before a potential agent has even been accepted for training by the CIA he has already been effectively brainwashed during the selection. Rees' 'leaderless groups' were, in reality, groups of candidates artfully manipulated by outside programmers in completely contrived and controlled situations. In 1946, Nathan Kline who at the time served at War Shipping Administration of the United States, described an officer selection process that Rees personally set up for the U.S. Marines shortly after the war."

Peter Cuskie in *The Shaping of the Anglo-American SS by War* explains, "Twenty candidates were gathered together in a group and told their future as a squad in the Marines was dependent upon beating the record of all other marine units who had tried to solve the problem they were about to be presented. Then, they were told to imagine they were on a deserted island and the unassembled life raft before them had just drifted ashore. With suitable appeals to their 'team spirit' they were instructed to break the record in getting the raft assembled and off the island.

"Psychological warfare specialists standing on the sidelines observed carefully the approaches of each individual in the group to the problem. Did he immediately rush into trial and error mode of operation or stand back and size up an overall solution? Did he exhibit gung-ho enthusiasm and motivation or withdraw in alienation from the situation? Which man would move for leadership and manage to enforce group discipline and draw out 'team spirit'?

"When they had tabbed their 'leader' a ruse was used, such as an alleged invasion on the other side of the island, to pull him out together with three or four others. This way, a new situation was developed where the psychological warfare experts could observe the rise of a new 'team leader.'"[33]

One purpose of these insidious and artificial schemes was to encourage mindless 'team spirit' and to select the most rabid and competent 'team leaders.' Another aim, when combined with the candidate's Personal History questionnaire and other written tests, was to compile a 'psychological profile' of each man for later use.

Again, citing Peter Cuskie: "It was still necessary, however, to destroy whatever real ego strength the candidate still possessed. This was the intention behind Rees' stress tests. One such test was used by John Gardner at the OSS... In this brainwashing scheme the candidate was given twelve minutes to construct a 'cover story' to be presented to interrogators who 'caught him' stealing government documents marked SECRET from a Government office in Washington. The candidate was told that this was a make or break test encouraged to internalize a plausible new identity, and warned that his answers to his interrogators must not compromise OSS organisational security or blow his cover.

"When the candidate was finished thinking up his cover story, he was brought into a dark room, a spotlight blinding him and three agents facing him. For the next several minutes these brainwashing specialists would almost always tear to shreds the victim's hastily concocted cover story by various 'tough cop-soft cop' methods, well-prepared staccato catch questions, physical brutality, etc.

Almost without exception, the candidate was left dazed and confused. Then the brainwashing agents would abruptly break off the interrogation saying: 'We now have abundant evidence that you have not been telling the truth. That's all.' After the interrogation board had ostentatiously engaged in some whispering back and forth would come: 'Your name is Jones, isn´t?...It is our decision, Jones, that you have failed this test.'

"The crest fallen candidate was instructed to go upstairs. There a staff member would pretend to commiserate emotions and fear in this pleasant atmosphere following the earlier great tension. Most of the broken candidates would readily open up and talk about their childhood in response to questions like 'As a psychologist I´ve been wondering whether there weren´t times in your childhood somewhat similar to this – when you concealed petty things from your mother when she questioned you.' Usually, the candidate would naively and pathetically ramble on about his mother, his early sexual experiences, etc. By this time OSS not only had its desired psychological profile – it also had crushed the last vestiges of genuine ego strength in its victimized potential agent and was now in a position to manipulate and program him almost at will to co-operate with 'the team.'

"The most up-to-date utilisation of these Anglo-American SS brainwashing methods was unwittingly revealed by the *Sunday Times* of London on January 27, in an article about the U.S. Army Special Forces entitled the "New Secret Service." The *Times* describes the counter-interrogation techniques now used.

"The four-stage brainwashing program starts with testing for sensitivity of the nervous system through flashing words or symbols on a screen for fractions of a second – in order to find if physical torture or isolation in solitary confinement would better 'break' him; phase two is the breakdown of troops' identity by teaching him how to develop a 'plausible alter ego' (alibi); the third phase involved brutal group attack or self-criticism sessions used to further destroy ego strength to supposedly 'simulate' enemy interrogation; and the final phase instruction in beating lie detector involved amongst other things pairing strong electric shocks with ordinary everyday words. This technique which would have obvious value for the CIA in preparing brainwashed veterans for hit squads is described under its counter-interrogation cover as follows: These ordinary words acquire an emotional meaning for the man trained for a certain mission. Thus, in an interrogation, with a mixture of control and conditioning, the man's response will appear chaotic and misleading."[34]

The OSS/CIA wasn't the only one using the latest and most up-to-date systems and techniques of counter-interrogation and agent indoctrination. Soon after, Tavistock developed newer, more and more sophisticated methods of mind control. Tavistock used various techniques of coercion, hypnosis, and the use of mind-altering drugs to achieve brainwashing for their victims, "all of which applied the same basic format: Induce massive physical or psychological stress in an individual, and then relieve that stress."[35] Through repeated vacillations between stress and relief, the subjects, be they army recruits, intelligence agents or general population eventually became intensely suggestible.

The OSS "pristine" interrogation "dark room" literally became the house of horrors for the newly selected recruits of the British MI6. One such secret location reserved for 'Special Government Projects' was at Powergen in Solihull, UK.

The year is 1979. "The young British Intelligence recruits entering this building had no idea of the hell that lay before them,"[36] writes Richard Tomlinson, former MI6 spy who was part of the program. "No idea, that they would be forced to become slaves to a demonic, mind control program run by MI6 and sanctioned by Royal Arch Freemasonry through Tavistock Institute."[37]

It was at Powergen in 1979, that a 21-year-old recruit, Richard Tomlinson, attending his first INSET British Intelligence, and was first shown a passport photograph of Vladimir Putin, supplied to him by one of Oleg Gordievsky's contacts. Gordievsky, the highest ranking KGB officer to escape to the West, was a Royal Arch Freemason.

The course was run by Stella Rimington (MI5), and John Scarlett, the overall MI6 controller of the modules. The two Tavistock designed programs for British Intelligence Graduate Trainees were:

1. The 'MI6 Beast' programming.
2. 'Smelly Cheeses' programming in conjunction with 'Jonathan Livinsgstone Seagull.'

The first one refers to the 'beast computer' program, "placed into every recruit's mind control programming during recruitment training at both MI5 and MI6. One of the tasks set the graduate trainees and mind-control programmers who were running the Royal Arch Freemasonry course, was a 'treasure hunt.' The treasure hunt was to test their 'spycraft' skills and general sadism. Part of this course was based on an old Roman game, which the soldiers had played with their prisoners. The Rite of Saturn. In short, it was a sado-masochistic game, whereby the recruits were tortured and abused in a similar fashion to the Roman prisoners. This was called the 'Via Dolorosa.'"[38]

"Via Dolorosa was a Tavistock-designed water torture – the programmers in British Intelligence bring you to a near-death experience in order to make you obey. It is also a way of separating the weak from the strong. The programmers torture the recruits repeatedly, in order to break their spirit. The key, an initiate is worthless to them if they are capable of disobeying orders from above. Some people never recover from the experience. Symbolically, this is the birth, death and resurrection process. The stages of the Cross; most initiates are then taken to Jerusalem to walk the Via Dolorosa in order to reinforce this programming, they have become 'Christ-like.'"

Smelly Cheeses is programmed in via the reading and enactment of *Three Men in a Boat*, a Jerome K. Jerome novel published in 1889. "Chapter 4 of Jerome's novel relates how it is most important whilst traveling and carrying 'smelly cheeses', not to touch the goods; nor afterwards, once one has arrived at one's destination! All smugglers are most worried about the safety of their drugs/contraband and therefore the importance of this part of the programming."[39]

In his self-published, unofficial biography, *The Golden Chain*, Tomlinson relates that "the other graduate trainee programmers were given huge electric shocks after 'training' i.e. the enactment of this little drama, and would then be instructed to sit on chairs facing a large window. This was 'relaxation' to de-stress the brain. The young men would sit there and grind their teeth (normal response to ECT) whilst gazing blankly at the sky. 'Mr Blue Sky' by ELO was sometimes played at this point, to 'rejuvenate' their spirits."[40]

"The main point behind MI6 'Smelly Cheeses' programming was that prospective recruits to British Intelligence were sent on 'gap years' around the world on various 'missions' i.e. to courier contraband on British-dominated smuggling rings, which had their origins in the British Empire's 'snake-lines,'[41] in clear reference to the Asian drug routes and especially African diamond routes.

'Smelley Cheeses' programming was used in parallel with 'Jonathan Livingstone Seagull.' Why? As Tomlinson writes, "to encourage and train those who had been picked to 'remote view' and then come back like 'homing pigeons' when called by the programmer."[42]

Fast forward to late spring 1980: Russian Monastery Compound, Ein Kerem, Jerusalem property of Russian Orthodox Church. Back in 1980, it was run as a nunnery but was known to be a KGB 'spy-center.' According to Tomlinson, "Vladimir Putin and the other KGB Royal Arch Freemasons, stayed within this high-security compound. The British Intelligence recruits mainly stayed at the 'Youth Hostel' behind the St John Church in the center of Ein Kerem. It is located partly up the mountain which overshadows the town and it backs onto a pine forest."[43]

What is mind-boggling in this upside-down universe of smoke and mirrors is that, "In the evenings, the 'graduates' would sit on the verandah of these buildings with their KGB Russian counterparts; to smoke, drink vodka and shoot the breeze."[44] Tomlinson spoke Russian and acted as a de-facto translator for the group.

Tomlinson explains that this was the first time that the future President of Russia, Vladimir Putin and the now former MI6 agent had met.

Fast forward to 1993. Poland. With the Soviet Union disintegrating into oblivion, Poland became the world's largest flea market in the world of intelligence. Possessing large mineral wealth, and large oil reserves, plus more timber than the Amazon, not to mention an immense stockpile of Soviet era weapons, the country was being asset-stripped. The strategy was intended to topple Russia into anarchy, towards the point where Russia could not oppose Western military operations.

"No one really knew who was working for which agency or which agency was allied to whom, in what turned out to be a 'smash and grab' gun fight. One could name MI5, MI6, the CIA and the IRA as but a few in this free-for-all; the desperate dash to acquire information, drugs, guns – you name it. Just about every intelligence agency and mafia organization was in on the game."[45]

Everything was on sale. Truckloads of Soviet rubles motored down autobahns. Many were used in complex swap operations in which billions of narco-dollars were laundered on behalf of the Ndrangheta, the Calabrian Mafia. Numerous prime western banks such as Bank of New York, Goldman Sachs, Fleet Financial and Bank of Boston looted up to $500 million.

Tomlinson explains: "Vladimir Putin had lost his job as KGB controller of the Stasi in East Berlin. There was a famine in St. Petersburg and the KGB owed their employees several months of salaries. Whole families were starving in the cities and Vladimir Putin's own family was no exception to this rule."[46]

Putin needed help and reached out to the very people who supposedly were his enemies, such as John Scarlett, MI6 Moscow Station. Like Putin, Scarlett was Royal Arch Freemasons.

Tomlinson is unable to shed any light on the deal that Scarlett struck with Putin but he does add several nuggets of information.

Putin, according to Tomlinson, "had been under MI6/Royal Arch Freemasonry mind control since 1979 and had been one of their operatives within the KGB until 1994. Putin needed safe passage for himself and his family out of a situation in St. Petersburg, which was rapidly deteriorating. Scarlett had agreed to set up a new identity for him in the UK as a teacher of German,"[47] as Putin spoke German fluently.

Putin's two daughters and his wife would follow him at a later date. However, somewhere along the line, Scarlett reneged on the deal. Instead of a safe passage out of Russia, MI5, for reasons unknown, had decided to torture and then murder Putin as soon as he set foot on British soil.

But British MI5 underestimated future Russian President, Vladimir Putin. Through his extensive network of contacts, he learned of MI5's real agenda and called the deal off, returning to Russia instead.

In 1994, Putin struck back. This year MI6 laptops began to disappear in Europe. Putin and his associates were tracking down the MI6 Senior Officers who held the color-coded information of the tailored mind-control programs for their officers and agents, within their laptops. Tomlinson explains that "once the codes are broken, it is relatively simple to control any of the operatives listed as well as guess how to control others who were 'indoctrinated' in a similar way and then to turn them against their 'masters.'"[48]

The main clue was to be found on an antique British typewriter '55 Imperial 55' at the Jerusalem prison museum. The code on the typewriter keys, runs as follows:

1. $SV
2. XDM
3. KGB
4. RKK
5. AAN
6. TYM
7. ZKTZ
8. NLT
9. SKTZ
10. PF!

The 10 sequences above accounts for 10 different branches of Royal Arch Freemasonry-infiltrated British Intelligence departments "from its inception (MI1) to the current, public form of MI5 and MI6."[49] For example, clue 7 is 'zygote' (Zayin, Chet, Tzadik, in Hebrew) which stands for the Royal Genoma Project and clue 9 is sex kittens. These departments are all still in existence today.

The KGB Royal Arch Freemasons were one of these departments, in this tarantula-like Masonic organization.

The CIA wasn't far behind the British MI6 in their approach to brainwashing. The Agency realized that the more maverick approach involved the "role of social factors in skewing the minds of the agents to the Agency's needs. In other words, individual brainwashing tactics, such as electroshock therapy or the use of drugs, though powerful, were no match for the power of suggestion in enforcing behavior."[50]

So Rees and his partners explored the deepest recesses of the human mind, adopting the object-relations approach of universal literature which emphasized imagery and symbolism rather than instinctual drives and psychic energy. Through the use of associated theory, the idea was to recreate the imaginary world of an immortal piece of literature in the minds of the test subjects, using the work of art for immoral purposes. Under Rees and Tavistock, these techniques became a testing ground for new forms of institutionalized psychological control.

THE BLUE BIRD OF HAPPINESS

The 'Land of Memory' has always been the primary objective of mind control and counter insurgency operations. There is a phrase which is perhaps not used so much these days as it was in the tender years of the twentieth century: "the blue bird of happiness." What many people do not realize – and did not realize even then – was that this term had its origins in *The Blue Bird* (1909), the most famous work by the Belgian Nobel Prize-winning author and dramatist, Maurice Maeterlinck. In this play, two children set off on a search for the Blue Bird of Happiness. This search leads them on many adventures – a kind of iniatic quest for the Holy Grail. Peter Levenda, in his brilliant book *Sinister Forces* explains that "many of the motifs of Maeterlinck's play are repeated in the CIA's search for perfecting mind control,"[51] a search that began with Project Bluebird.

The story, "which begins on Christmas Eve, involved two children – Tyltyl and his younger sister Mytyl – who set out on a quest to find the Blue Bird of Happiness. Impoverished children of a woodcutter, who live across from a great house with very rich children, they understand that they are too poor to receive Christmas presents that year. In the middle of the night, there is a knock on the door, and an old woman – who later introduces herself as the Fairy Berylune – asks them if they have "the grass that sings, or the bird that is blue." Berylune has a sick daughter who will not get well unless the Blue Bird of Happiness is found. The children, eager to help, set off on a quest for the mysterious Bird... and to visit their dead grandparents, with the Fairy's help. In order to visit the dead, however, they have to pass through the Land of Memory, which is on the way to the Blue Bird."[52]

I am taking my time to explain Maeterlinck's work in order to help the reader understand how secret societies, intelligence agencies and government use literature to indoctrinate their cadres into becoming mind control slaves to the elites. Blue Bird is one such program devised by Tavistock specifically for the CIA.

Just as in Maeterlinck's play, the recruits "are given a magic hat. On this hat is a diamond, set squarely in the center. When Tyltyl presses it, he will see "the Soul of Things"; if he turns it to his right, he sees the past; to the left, the future; and as long as he wears the hat, it is invisible. Those who read Eastern mysticism will no doubt grasp at once that the diamond is the 'adamantine substance' and its position on the hat is a reference to the Third Eye, which, when opened, gives the devotee access to secret information and occult powers. The search for the Blue Bird will give Tyltyl (and mind control recruits) these powers as they walk through the Land of Memory, the Palace of the Night, a Graveyard, and an enchanted Forest. Eventually, of course, the children arrive back at their home on Christmas morning where they discover that the Blue Bird of Happiness has been there all along."[53]

However, for Tavistock, the play is full of occult and esoteric elements of necromancy treasure-quest. That children – virgins – are the ideal seers according to the occult texts of the Middle Ages goes unmentioned. The moral is uplifting and spiritual, conservative and charming. A fairy tale replete with talking animals and trees and kindly grandparents. The quest itself, though, is enlightening for other reasons.

The Land of Memory, of course, was the target of the BlueBird project: to enter that Land in another person's mind, to go through the drawers, rearranging the furniture, and leave unnoticed. "Once the Korean War started, and American POWs began making bizarre, pro-Communist statements after a mysterious sojourn in Manchuria, the world was introduced to the concept of 'brainwashing', and the BlueBird took on enormous importance. If the Communists could alter the consciousness of American soldiers, then the War took on a completely different nature: it became a war of culture against culture, of atheism against religion, of race against race, of Darkness against Light."[54] This was a war not to be fought by bullets alone; psychological warfare operations were ramped up at the same time as BlueBird went into full swing, and what William Sargant would call in 1957 'The Battle for the Mind' had began.

Tavistock had obviously known of Maeterlinck's creation, just as Lewin and Rees had known of his other, more obscure works or of his studies of astrology, psychic phenomenon and mysticism. Maesterlinck's description of the Land of Memory had to strike deep resonances in the minds of the Tavistock team as they embarked on their search for a key to the mysteries of human consciousness. And so, the mind control programming begins.

"In the Land of Memory – a strange land enshrouded in mist and darkness – the two seekers find their grandparents, long deceased. Their grandparents tell them, 'The last time you were here, let me see, when was it? ... It was on All Hallow's, when the church bells were ringing.'[55]

"All Hallow's, of course, is the Day of the Dead. They find a Blue Bird, but when they return from the Land of Memory the bird has turned completely black. Only the first 'trial' has taken place, and Tyltyl and Mytyl have much farther to go.

"Next stage in mind control is The Palace of the Night. It is a somewhat more eerie and forbidding place than even the Land of Memory. Night is depicted as a kind of angel, a Dark Lady with wings in place of arms. The Palace of the Night is the domain of Ghosts, Sickness and Wars. One goes directly from the Land of Memory to the Palace of the Night, from unlocking the secrets of the memory to communing with the dead, controlling sickness and becoming victorious in war. Blue Bird, and its associate programs such as MK-ULTRA and OFTEN, became involved in all aspects of behavior modification, hypnosis, drug-induced psychological states and the creation of amnesia.

"Tyltyl finds a great door set in the rear of the Palace of the Night, and is told that he must never open it, the great dangers await those who rush in and that no one who enters that room ever returns to the land of the living. Tyltyl, who is charged with the sacred quest of finding the Blue Bird, finally decides to open the door. When he does so, he sees a beautiful garden, a waterfall, many wonderful things and ... of blue birds. He tries to grab as many as he can, and when he takes them out of the room they turn corpses in his arms. They cannot stand the light of day."[56]

The sessions continue. "Both, the brainwashing trainee and Tyltyl must continue their quest, leaving the Palace of the Night for their next destination, the Palace of Luxury. Here, they are confronted by a scene of outrageous depravity, as fat beings eat, drink, laugh and carouse, beatific in their willful ignorance. As always, Tyltyl – with his magic diamond (secret knowledge not available to others) – can see things as they really are, and by pressing the diamond he perceives that these beings are miserable fools and thus, exposed, they retreat to the

Miseries, a special place from whence they may never return. Of course, the CIA specializes in "seeing things as they really are," looking behind the curtain, behind the facade, and their entire program may be seen as Tyltyl's diamond: they put the magic diamond to use to see the human mind as it really is, to strip away a human being's conscious defenses so that everything, every secret – even highly classified government secrets that could cost the lives of many – would be revealed.

"From the Luxury Palace, Tyltyl proceeds to the Kingdom of the Future. This is where he finds a land filled with children who are not yet born, dressed completely in blue and like little scientists. They are occupied with the inventions that they will produce once they have been born on earth. However, Tyltyl – as a living person – is not allowed in this Kingdom, which is under the rulership of Time. From the Kingdom of the Future, then, to the Graveyard.

"This is much like any other cemetery, with tombstones and grass and silence. At midnight, he is to use his magic diamond once again to see the Dead. In the midst of the darkness and impending horror, a clock in the distance chimes the hour. Tyltyl presses his diamond, frightened but willing to face the Dead as the next in his series of trials before he can capture the Blue Bird. But instead of ghostly figures in shroud and clanking chains, the scene changes. The tombs open, but instead of ghoulish entities only flowers emerge.

"They had thought that ugly skeletons would rise from the earth and run after them. They had imagined all sorts of terrible things. And then, in the presence of the Truth, they saw that all they had been told was a story and that death does not exist.[57]

The motto of the CIA, of course: 'You shall know the Truth, and the Truth will set you free.'

From the Graveyard, Tyltyl wanders into the Forest. During the quest, he has been accompanied by various other creatures who have been observing Tyltyl from a distance. Many of these creatures – most especially the Cat – fear for their lives should Tyltyl be successful. The Cat says: "It is better to rely on one's self alone. In my cat-life, all our training is founded on suspicion. I can see that it is just the same in the life of men. Those who confide in others are only betrayed. It is better to keep silent and to be treacherous one's self."[58]

This might well have been the personal motto of some of the sinister men who worked behind the scenes at Tavistock and the CIA, men like John Rawlings Rees, Eric Trist, CIA director Richard Helms and the MK-ULTRA czar, Dr. Sidney Gottlieb: Loners, desperately private men who guarded the secrets of the nation and the secrets of their lives with equal passion, since the two were so often inextricably intertwined, their training certainly founded on suspicion.

In the story, the Cat leads Tyltyl into a trap in the forest, where he is set upon by the trees and the animals and has to fight for his life. He is saved at the last minute by Light, who warns him that "man is alone against all the world."[59]

Finally, the Leave-Taking occurs, as the hour has grown late and the children return to their cottage miraculously as the clock strikes eight on Christmas morning. This is the Awakening, a constant theme for the CIA and Tavistock. Of course, when they awaken, the children realize that the Blue Bird has been in their home all along.

Through an "innocent" child's tale, Tavistock brainwashers have embarked on a sacred quest that would lead them into humanity's deepest secrets; by delving into the universal, macrocosmic secrets of the human mind, they hoped to uncover the specific, microcosmic secrets of their enemies. As above, so below.

Tavistock used their understanding of psychiatric methods to formulate and implement an action program based upon such beliefs. Once the neurotic map of each individual was determined, Tavistock was able to set up a "filtering" mechanism, that is different forms of brainwashing, to select various neurotic types and place them in their appropriate settings.

There is deeper symbolism involved in all of these Tavistockian processes. Tyltyl and Mytyl were innocent, virgin children, morally spotless, on an initiatic journey in the company of Light. The men of Blue Bird and later MK-ULTRA, could hardly be considered in the same way. From the point of view of the ancient mystery religions whose quest they were imitating, they were plumbing the depths of the Abyss without having undergone the period of purification; thus their sins – their personal, private, specific sins – would come to haunt them in the days and weeks and years that would follow, soiling their reputations and forbidding them entry into the Inner Temple.

*****

The psychopathic kernel of Tavistock's long-range vision is therefore this. In *Campaigner* magazine's ICLC Strategic Studies, L. Marcus explains, "The fascist 'structural reforms', local community control and 'social contract', are the assertion of the unconscious infantile realm at the expense of the relative rationality of former socialising ego-ideals. Fascism is the world desired in the paranoid dreams of the "Id." Conversely, if the "automized individual's world is converted into a controlled environment which conforms to such fascist 'structural reforms', the victim's mind will discover that only its potential paranoid self provides it with the means for agreement with that controlled environment."[60]

On one level, the technique was applied to the world of intelligence, but a far more hideous use was envisioned for it by the practitioners of the art of brainwashing. After the war, these techniques were put back into society and applied to real lives and real people. Tavistock had understood that the family was the most powerful psycho-active agent. One of their creations, "therapeutic groups," which we will analyse in this section, provided the opportunity to tap the power of the family.

The most advanced among the brutal practitioners of the new industrial psychology was Dr. John Rawlings Rees. Rees discovered that an unreal realm could be created: the social group. An individual is made to transfer his or her identity to the group, wherein they become subjected to the most intense forms of suggestion. Provided the individual's inner sense of real identity is destroyed, he can be manipulated like a child.

M. Minnicino in an article titled "Low Intensity Operations" explains: "A skilled group leader can use the group to create a powerful, albeit artificial, "family" environment. Once this environment is induced, a therapist can manipulate a member of the group, not by direct attack, but by subtly manipulating the other members of the group, for example through suggestion. If the victim has been sucked into thinking of the group as something helpful (motherly), then, when that environment has been manipulated to turn against him, it will tend to have the impact of deep motherly rejection."[61]

Rees and Tavistock understood, through their extensive engagement in group work, that manipulation based upon bourgeois conceptions of the outside world as magical (mother's fears) was fundamental to mass control. In other words, the Institute looked for methods by which to manage populations by exploiting their ideology. By first creating numerous social groups and then setting such groups into competition, and making all gains contingent on winning these at the expense of the contingency groups, a self-policing fascist social order could be established. It was merely necessary to atomize the subject population, employing an "armamentarium of sociological and psychiatric weaponry to the effect of so dividing the people against itself,"[62] setting race against race, language group against language group, women's 'self-interest' against men as 'oppressors,' further subdivide them by professional category and so on, and to further subdivide them into small territorial community 'self-interest' groups, that the military forces are "never confronted with more than a small material force represented by a hard core resistant to a generally effective combination of sociological and psychiatric weapons of control."[63]

One of the ways in which such practices were encouraged and used by Tavistock brainwashers involved increasing levels of productivity and the intensification of labor at the expense of one's health. In other words, Rockefeller induced self-motivated destruction of the self, a society of brainwashed zombies, content to subsist on the brink of starvation, performing sadomasochistic acts of sodomy for a perceptibly irrational sort of psychotic holocaust. All of it is based on a sort of perverse dionysian sodomic mindless fanaticism within the broader framework of Rees's Rockefeller model of fascist society. "The purpose is to function under a spiralling reduction in standard income and working conditions."[64]

There is a method to this madness, however.

For example, in an environment of downward spiralling standard income and working conditions, higher and higher productivity is "suggested" as the group's goal, often at the expense of safety and the physiological well being of the group members. Anyone who protests is told they are maladjusted. "The idea was to re-create a family dynamic, or a dynamic of peer pressure, in group therapy, where predetermined objectives were forced onto the group through consensus, or 'democracy,' in the language of these social engineers. The idea was that by attacking someone's sovereign identity in the group, that individual would forfeit their sovereignty to the group and become suggestible to the predetermined objectives."[65] Such techniques play upon the "vulnerable person's guilt, shame and regret with both sledgehammer and scalpel."[66] Humiliated by the group, with only the unreal realm to judge from, the individual further degrades himself and accepts the verdict. He begins producing more.

This is the malicious, disgusting kernel of co-participation, "quality of life," co-determination, "humanization," human relations or whatever other euphemism may be mustered, all under the banner of 'Post Industrial Society,' a Ford Foundation-sponsored piece of garbage whose particular point, as a counterinsurgency weapon of ideology, introduced the proto-fascist notion of 'Zero Growth.'

As M. Minnicino explains, "The slave labor and death camp system of the Nazis are not quirks of Hitler and his associates, but an intrinsic expression of the fundamental policies of any 'Zero Growth' economy. It is impossible to have Zero Growth policies beginning today without mass genocide tomorrow."[67]

In fact, the idea for these programs came long before Tavistock was established. John D. Rockefeller Jr. began to push the savage "human relations" plans as the best way to get more productivity in 1916 when he spoke at the YMCA Industrial Conference and then again in 1917, during his speech at Cornell University Founder's Day convention. One of the key proposals coming from the Rockefeller camp was to push a "democratic" employee stock ownership plan, because "it makes the worker a capitalist in his viewpoint and this renders him a conservative and immune from radical ideals," and produces greater productivity, which is the aim.

The 1940's was the turning point for the Rockefeller strategy of behaviour modification-brainwashing, co-determination, co-participation corporativism for the takeover of the United States and world labor movement.[68] As war wound down, Rockefeller changed the psychology of the workers in ways essential to the way he would rule the U.S. organized labor movement from then on. This was done simultaneously on several levels.

In 1946, Rees-Tavistock-Rockefeller was formalised in the Rockefeller Memorandum, in which Brigadier General John Rawlings Rees exposed the full depth of his "cabal and offered it to the family. The Rockefeller Foundation, which had been heavily funding

Tavistock and Tavistock members since 1934, readily accepted, and the Tavistock Clinic was transformed into the Tavistock Institute of Human Relations."[69] The Institute developed national and international operations under the heading "operation research."

Firstly, Rockefeller took many of the social scientists who had been involved in the bestial-fascist war intelligence services and set them up on campuses in Labor Institutes funded with Rockefeller Foundation, National Institute of Mental Health, Army, Navy, Air Force and large capitalist firm money, guidance and projects. These Labor Institutes were created at the same time as the CIA, the Joint Chiefs of Staff, and the National Security Council as part of the same network that Rockefeller was building to rule the world, once the United States had replaced Britain after the war as the world strongest power.

One of the key Tavistock projects was Rockefeller-sponsored Labor Institute, where behaviour modification, inner group motivation, team-working, social dynamics, productivity increasing, were all studied from the psychological point of view to manipulate the great unwashed.

By the early 1950s Rockefeller had a network of labor institutes plus control of the Department of Labor to physically and psychologically control from top down the United States labor movement.

But Rockefeller's greatest contribution to the "labor cause" has been the complete destruction of the labor movement, its perversion and control through Rockefeller-Tavistock orchestrated operations. From Civilian Conservation Corps (CCC) to Works Progress Administration (WPA), from National Civic Federation (NCF) and American Federation of Labor (AFL) were labor movements in the United States always firmly under the control of the Rockefeller financial interests. For example, Samuel Gompers, the AFL's first president, helped form in the early 1900's the National Civic Federation. "It believed in the supremacy of employers and the benevolence of capitalism and helped break strikes, recruit vigilantes, and conduct attacks against working class. During the First World War, Gompers championed the vicious War Labor Board that crushed labor. Among the NCF's leaders were Mark Hanna of the U.S. Steel Corporation and such Rockefeller followers as Charlie Elliott, member of the Board of Trustees of the Rockefeller Foundation and president of Harvard."[70] Another of the Rockefeller organizations was Steel Workers Organizing Committee, whose director was Philip Murray. Murray's right hand man was Clinton Golden, who in 1947 joined the editorial board of Tavistock's *Human Relations* magazine and in the mid-1950s became a member of the Rockefeller-controlled Ford Foundation.

Yet another Rockefeller front was a Cornell ILR School created Coalition of Labor Union Women, which held its first conference in Chicago in 1974. The group was intended to push Affirmative Action for women, pitting men against women in vicious job fights with millions being laid off.

"The three Rockefeller-formed and funded and directed labor institutes, Tavistock, the ISR and Cornell ILR, are only three of the more than two to three hundred that were started by the Rockefellers throughout the world after the Second World War. They include the Center d'Etudes de Problemes Humaines de Travail and the Association pour la Recherche et L'Intervention Psycho-sociologiques in France; the Industrial Research Institute in Bonn, Germany and the DGB's (German Trade Union) Department of Social Relations Institute, also in Germany; la Sociedad Argentina de Investigación Operativa (SADIO); the International Jewish Research Foundation on Human Relations in Israel, and the Psychological Institute at Kyushu in Japan, to name a few."[71]

The list[72] of Rockefeller influence is enormous. "Between the Rockefeller family's long-standing control over the American Medical Association and American Psychiatric Association, and the encroachment of the CIA establishment into all branches of government, various governmental, including military funds, were used to push the development of brainwashing and to plant Rockefeller-Tavistock proteges in key positions and institutions. For example, B.F. Skinner's position at Harvard was established in this way. Another Rees student, Dr. Kenneth Clark, moved into Rockefeller's New York State Board of Regents and also into the key black counterinsurgency post (MARC) of the Rockefeller-inspired and Rockefeller-controlled Ford Foundation. Dr. Nathan S. Kline, one of the worst Reesian criminals, headed New York State's Rockland State Hospital as well as held a key position at New York's Columbia Presbyterian Hospital, where a broad and old tradition of brainwashing thrived."[73]

In fact, the techniques used in labor negotiations in Spain, France, Germany, the United States and most other western countries come straight out of the Tavistock role-playing manual. It would do labor leaders some good to study Rockefeller-Tavistock techniques for labor negotiations. Or perhaps, with a wink and a nod, they already do. Is that why the Spanish labor movement is a bad joke? Let's have a look.

"First even before a strike a union is thoroughly surveyed. Rockefeller Labor Institutes do psychological profiles on the union as a whole, collecting information on it through students who are sent out with questionnaire, by counselling done on union members, by intensive questioning of union leaders, by attending set up discussion sessions of union members and attending union meetings, by evaluating the union's past history such as its tendency to strike, for example, and by going over written tests and records of personal histories of a significant number of individual members. Studies are done on racial and ethnic subsections of the union. For every subsection, such as Italians, for example, it is recorded whether they are from middle income, lower working class, newly immigrated, perhaps without citizenship papers, deeply religious, for a long time from a certain neighbourhood, very attached to one's mother and so forth. Then the union and subsections are evaluated for how they react under crisis situations: the pitiable psychological weaknesses, the neurotic guilts, the sort of horrifying images that produce fears in members, which will then be played on in the media, and subtly exploited on unconscious levels in government and other Rockefeller propaganda, such as the known fear of white middle income teachers of attack from black gang youth, the points at which members crack under stress, the methods to make them hysterically panic, at what point members will stop trusting each other, their intimidation by violence, what vicious outside pressure will make them cave in and so on. As many as 100 or 200 studies on every union are done by Rockefeller Labor Institute."[74]

The methodology employed come straight out of a *Behaviour Theory of Labor Negotiations*, written by Walton and McKenzie for the Rockefeller-created Cornell School of Industrial Relations. First step: The union is provoked by a particularly low and insulting contract offered. The furious union comes out en masse to the picket lines. Clenched fists and determined, stern faces are everywhere. Rockefeller waits out the first few days. By the afternoon of the third day, the union leader is spending most of his time in the union hall, chatting with his friends. By the afternoon of the fourth day, the shouts are not half as energetic and few, if any faces hold the stern, determined look of the first morning.

Tavistock enters the scene. The strike is to be crushed in stages. "The union leadership is called into negotiations. Complete psychological profiles have been done on the personality structure, etc. of the union leaders. In meetings, structured like a group therapy session, unknown to the union leaders, the behavioural modifier arbiter conducts psychological probes against the union leaders. Most union leaders are schlemiels and can be easily manipulated. The negotiators and the capitalists know that the pressure of a media attacks, the tensions of the strike, the pressure from the rank and file, and the expiring strike fund are operating on the leaders, making them internally wrecked.

"Then, the leader is brought in for further negotiations. At this point, they are close to begging for anything to take back to the membership."[75]

But, instead of better conditions for the workers, the union leaders are shown the techniques to sell it to the membership so that they won´t think they are being cheated. How is it done? The union leaders and the shop stewards are given a course on negotiations at the nearby Labor Institute. Here behaviour modification is performed. Guess who runs the seminars? You guessed it, loyal Rockefeller psychologists and their staff members.

However, the union penetration and eventual take-over doesn´t end there. Key agents, behaviour modifiers, are placed into union positions, usually in the posts of educational secretary and attorney.

"By the next contract negotiation session, the union leaders are more easily malleable and will often agree to contracts in pre-contract negotiation therapy sessions. Furthermore, attention is simultaneously turned to the membership. Psychological warfare techniques such as 'cooling off periods' between negotiations, smear campaign in the media, offers of conciliation, are used and are objective forces Rockefeller has at his disposal to break the remaining will of the strikers to crush them."[76]

WARTIME DEVELOPMENT OF BRAINWASHING

In practical terms, Rockefeller and Tavistock technocrats used their knowledge to break and then to control a powerful Coal Miners Union, who according to their own propaganda, "occupied a position of unquestioned leadership in the history of American labor."[77] The project director was a Rockefeller man, Eric Trist. Trist had written *Social Structure and Psychological Stress, in Stress and Psychiatric Disorder* – a piece of trash to be used in strikes and riots. With the post-war economy on a downslide and tensions amongst the trade unions running high, an alternative was devised.

Trist suggested that workers be appointed to the board of directors of major companies. What was thought at the time to have been a major victory for labor, turned out to be a brilliantly schemed plan of the Tavistock Institute. "The profitability of the company – whether it stays afloat and the worker has a job – rests on outside world economic developments and internal affairs. The outside matters, such as the global markets, are directed generally by commodity cartels' control of money, raw materials, and so on. The inside affairs unless there is investment in technological improvements, rests almost exclusively on the brutal speedup workers will subject themselves to at the expense of fatal accidents and death. The operation to place workers on the board of directors meant tying workers to murderous speedups to insure profitability. This came about by inducing workers to internalise this 'need,' creating a whip in their minds, and disarming other workers by pressure from their mates. In a depression, such schemes consummate in frantic, anxiety-ridden compulsion among workers."[78]

MARSHALL PLAN

The Rockefeller plans for post-war Europe were already being shaped during the Second World War. In 1946, Tavistock would become an important ally in the Rockefeller fight to take over labor unions around the world. One of the key events that shaped that fight was the Marshall Plan.

Named for the speech on June 5, 1947, by former General, and then-U.S. Secretary of State, George Marshall, who proposed a solution to the disintegrating economic and social conditions that faced Europeans in the aftermath of World War II. Under the program, the United States would provide "aid to prevent starvation in the major war areas, repair the devastation of those areas as quickly as possible, and invite European countries to join in a co-operative plan for economic reconstruction."[79]

What is less generally acknowledged is that the "Plan" came with strings attached. America made explicit requirements for trade liberalization and increases in productivity, thus "ensuring the Americanization of Europe as European political and economic elites became wedded to their American counterparts with no significant economic or political development taking place without U.S. approval."[80] Between the years 1948 and 1951, when the Marshall Plan was formally in operation, Congress appropriated $13.3 billion in aid to sixteen western European states. This credit was used to buy up European industry and the European working class extremely cheaply.

As Mike Peters writes in "Bilderberg Group and the Project of European Unification": "This unprecedented exercise of international generosity, dubbed by Churchill the 'most sordid act in history,' served direct economic purposes for the internationally oriented U.S. corporations which promoted it."[81]

Kai Bird described the hidden aspects of the Plan in his book, *The Color of Truth*, on the Bundy brothers. In 1949, "McGeorge Bundy, former Ford Foundation President took on a project with the Council on Foreign Relations to study Marshall Plan aid to Europe... The Council's study group included some of the foreign policy establishment's leading figures. Working with young Bundy were Allen Dulles (future CIA director), Dwight Eisenhower (future U.S. President) George Kennan (a key ideologue of the Cold War), Richard M. Bissell and Franklin A. Lindsay. Dulles, Bissell and Lindsay...would shortly become high-ranking officials of the newly formed Central Intelligence Agency.... Their meetings were considered so sensitive that the usual off-the-record transcript was not distributed to Council members. There was good reason for the secrecy. Participants were privy to the highly classified fact that there was a covert side to the Marshall Plan. Specifically, the CIA was tapping into the $200 million a year in local currency counterpart funds contributed by the recipients of Marshall Plan aid. These unvouchered monies were being used by the CIA to finance anti-Communist electoral activities in France and Italy and to support sympathetic journalists, labor union leaders and politicians."

On the one hand, the origins of the Marshall Plan are, in fact, to be found in the policy formation networks around the Rockefeller circles in 1942. "Key Rockefeller figures included Paul Hoffman, president of Studebaker Company. Hoffman was the leading economist for Committee for Economic Development, one of the Rockefeller controlled organizations. The CED devised some of the essential work for the Marshall Plan to loot Europe. Before this outright takeover began in 1947, a torturous psychological softening up was conducted against the European working class, including near starvation. In Germany, daily per capita consumption of calories fell to 1,300. Furthermore, the American Military Government that was occupying the U.S. zone cut off the supply of home heating fuel by diverting coal away from Germany. A systematic terror campaign began, with Reesian-trained death squads roaming Germany and murdering people... Part of the same operation was the selection procedure conducted by Reesian psychologists to weed out the most loyal and true to the Tavistock creed as the future rulers of Europe. Workers rendered desperate, capitalists starved for credits, rulers already chosen, the Marshall Plan machine for 'reconstruction,' that is outright looting, began in 1948.

"To guarantee that no trouble would come from the labor movement, the intelligence services, the CIA and the State Department along with Tavistock Institute, gathered key unionists to carry out the Rockefeller plan of reshaping and taking control over section of European labor movement."[82] These key unionists were heavily recruited from the National Labor Relations Board, a Rockefeller corporatist brainchild run by his loyal lapdog, Arthur Goldberg. "Goldberg was working with the social democrats within the European labor movement. He was funnelling OSS (CIA precursor) funds into the social democratic wing of the clandestine French federation of labor, preparing arrangements that would help U.S. intelligence take over the French movement after the war."[83]

On the other hand, the movement to form a united Europe was part of a larger plan to form a world government. The Hague

Congress, held in May 1948, "called for a United Europe and issued seven resolutions on aspects of political union. Number seven stated, the creation of a United Europe must be regarded as an essential step towards the creation of a United World... The Marshall Plan, aside from helping put Europe back on its feet, led to the Schuman Plan in 1950, when the French Foreign Minister Robert Schuman proposed that the entire coal and steel production of both France and Germany be placed under the authority of one supranational body,"[84] which in turn led to the European Coal and Steel Community, then to Euratom (the European Atomic energy Community), and ultimately to the Common Market.

ECSC was the first concrete step toward political unification, the first block of Empire building, the Empire being the One World Company Limited. With the signing of the Treaty of Rome, paving the way for the European Economic Community in 1957, the next step towards a future world company limited was taken.

*****

What I have described in this chapter, as horrendous and unbelievable as it may seem, is no myth. It is a hideous reality of the everyday encroachment of fascism with a democratic face upon our society devised at Tavistock Institute with financial help from the Rockefeller family going back over 100 years, to the beginning of the 20th century. This 'reality', which if faced as reality and not a figment of one's overzealous imagination, can be wiped off the face of the earth. However, if that reality is not dealt with up front, then within a relatively short period of time, everything else we do as people, every hope we have for a better world will come crashing down into a dustbin we call history. It is the Tavistock Institute or us, the people. There are no other alternatives.

Endnotes

1. Carol Menzel, "Coersive Psychology: Capitalism's Monster Science," *The Campaigner*, February-March 1974

2. Ibid

3. Ibid

4. Ibid

5. Carol Menzel, Coersive Psychology: Capitalism's Monster Science, *The Campaigner*, February-March 1974

6. Ibid

7. The Tavistock Grin, The Real CIA – The Rockefellers' Fascist Establishment, L. Marcus, *The Campaigner*, April 1974

8. Ibid

9. Rees, John, The Shaping of Psychiatry by War, Thomas William Salmon memorial lectures, Chapman and Hall, 1945

10. The Tavistock Grin, The Real CIA – The Rockefellers' Fascist Establishment, L. Marcus, *The Campaigner*, April 1974

11. Ibid

12. Ibid

13. *Sinister Forces – The Manson Secret: A Grimoire of American Political Witchcraft*, Peter Levenda, Trineday Press, June 2011

14. Low Intensity Operations: The Reesian Theory of War, M. Minnicino, *The Campaigner*, April 1974

15. Vivian Freyre, One Hundred Years Towards Incest, *The Campaigner*, February-March 74, Volume7. #4-5, p.62

16. Carol Menzel, Coersive Psychology: Capitalism's Monster Science, *The Campaigner* February-March 1974

17. K. Lewin 1942, "Time Perspective and Morale," in G. Watson, ed., *Civilian Morale*, second yearbook of the SPSSL, Boston: Houghton Mifflin

18. Carol Menzel, Coersive Psychology: Capitalism's Monster Science, *The Campaigner* February-March 1974

19. Rockefeller's Fascism with a Democratic Face, *The Campaigner*, Vol. 8, #1-2, November-December 1974, p.55

20. Ibid

21. Ibid

22. Low Intensity Operations: The Reesian Theory of War, M. Minnicino, *The Campaigner*, April 1974

23. Rockefeller's Fascism with a Democratic Face, *The Campaigner*, Vol. 8, #1-2, November-December 1974, p.56

24. Rockefeller's Fascism with a Democratic Face, *The Campaigner*, Vol. 8, #1-2, November-December 1974, http://www.campaigner-unbound.0catch.com/fascisms_rape_of_the_mind.htm

25. Ibid

26. Ibid

27. Ibid

28. Ibid

29. Ibid

30. Ibid

31. Ibid

32. Rockefeller's Fascism with a Democratic Face, *The Campaigner*, Vol. 8, #1-2, November-December 1974, p.44

33. The Tavistock Grin, The Shaping of the Anglo-American SS by War, Peter Cuskie, *The Campaigner*, May 1974, p.28

34. The Tavistock Grin, The Shaping of the Anglo-American SS by War, Peter Cuskie, *The Campaigner*, May 1974, p.29-30

35. INSNA: 'Handmaidens Of British Colonialism.' http://www.larouchepub.com/eiw/public/2007/eirv34n47-48-20071207/27-37_747-48.pdf

36. Richard Tomlinson and the Russians, http://www.richardtomlinsonandtherussians.blogspot.com.es/

37. Ibid

38. Ibid

39. http://www.richardtomlinsonmi6.blogspot.com/

40. One should be forewarned that there is no independent corroboration of these claims available; basically all we have are Tomlinson's recollections on this issue.

41. http://www.richardtomlinsonmi6.blogspot.com/

42. Ibid

43. Ibid

44. Ibid

45. *The Golden Chain*, Richard Tomlinson, self-published, 2005

46. *The Golden Chain*, Richard Tomlinson, self-published

47. http://www.richardtomlinsonmi6.blogspot.com/

48. Ibid

49. Ibid

50. INSNA: 'Handmaidens Of British Colonialism', David Christie, *EIR*, December 7, 2007, p.27

51. *Sinister Forces – The Nine, Book 1*, Peter Levenda, Trineday Press, May 2011

52. Ibid

53. Ibid

54. Ibid

55. Act II, Scene 2, The Land of Memory, *The Blue Bird*

56. *Sinister Forces – The Nine, Book 1*, Peter Levenda, Trineday Press, May 2011

57. Maurice Maeterlinck, *The Blue Bird: The Wonderful Adventures of Tyltyl and Mytyl in Search of Happiness*, Silver-Burdett and Company, p.113

58. Maurice Maeterlinck, *The Blue Bird: The Wonderful Adventures of Tyltyl and Mytyl in Search of Happiness*, Silver-Burdett and Company, p.4

59. Maurice Maeterlinck, *The Blue Bird: The Wonderful Adventures of Tyltyl and Mytyl in Search of Happiness*, Silver-Burdett and Company, p.128

60. Rockefeller's Fascism with a Democratic Face, *The Campaigner*, Vol. 8, #1-2, November-

December 1974, p.56

61. M. Minnicino, The Tavistock Grin, Low Intensity Operations: The Reesian Theory of War, *The Campaigner*, April 1974, p.39-40

62. M. Minnicino, The Tavistock Grin, Low Intensity Operations: The Reesian Theory of War, *The Campaigner*, April 1974, p.14

63. Ibid

64. Rockefeller's Fascism with a Democratic Face, *The Campaigner*, Vol. 8, #1-2, November-December 1974, p.63

65. David Christie, INSNA: Handmaidens of British Colonialism, *EIR*, December 7, 2007

66. Carol Menzel, Coersive Psychology: Capitalism's Monster Science, *The Campaigner*, February-March 1974

67. M. Minnicino, The Tavistock Grin, Low Intensity Operations: The Reesian Theory of War, *The Campaigner*, April 1974, p.16

68. Rockefeller's Fascism with a Democratic Face, *The Campaigner*, Vol. 8, #1-2, November-December 1974, p.63

69. M. Minnicino, The Tavistock Grin, Low Intensity Operations: The Reesian Theory of War, *The Campaigner*, April 1974

70. Rockefeller's Fascism with a Democratic Face, *The Campaigner*, Vol. 8, #1-2, November-December 1974

71. Rockefeller's Fascism with a Democratic Face, *The Campaigner*, Vol. 8, #1-2, November-December 1974, p.72

72. Komer, R. The Malaya emergency in Retrospect: organization of a successful Counterinsurgency Effort, *RAND Report R-957-ARPA*, February 1972, p.3

73. M. Minnicino, The Tavistock Grin, Low Intensity Operations: The Reesian Theory of War, *The Campaigner*, April 1974, p.23

74. Rockefeller's Fascism with a Democratic Face, *The Campaigner*, Vol. 8, #1-2, November-December 1974

75. Ibid

76. Ibid

77. http://www.umwa.org/?q=content/brief-history-umwa

78. Rockefeller's Fascism with a Democratic Face, *The Campaigner*, Vol. 8, #1-2, November-December 1974, p.65

79. "For European Recovery: The Fiftieth Anniversary of the Marshall Plan," Library of Congress, http://www.loc.gov/exhibits/marshall/marsh-overview.html

80. Richard Greeves, *Who rules the world?* Essay, self-published

81. Mike Peters, Bilderberg Group and the Project of European Unification, *Lobster Magazine*

82. Peter Cuskie, The Tavistock Grin, The Shaping of the Anglo-American SS by War, *The Campaigner*, May 1974, p.28

83. Peter Cuskie, The Tavistock Grin, The Shaping of the Anglo-American SS by War, *The Campaigner*, May 1974, p.73

84. Abolishing our nation – step by step," *The New American*, September 6, 2004

CHAPTER TWO

Tavistock and the Unholy Alliance

The religious and mystical dimensions to the story of Tavistock are central to any study of the US government's postwar interests in how psychology and parapsychology could benefit the intelligence agencies. It was Tavistock and the cabal of scientists who were the first to bring the potential uses of paranormal abilities in military applications, as well as the first in developing chemical substances that would stimulate psychic abilities. This cabal included men, such as Dr. Sidney Gottlieb, head of CIA's Technical Services Staff, with shadowy connections both to Operation Paperclip on the one side and to the Kennedy assassination on the other.

As Peter Levenda explains in *Sinister Forces*, "swirling about the feet and hands of the Kennedy assassination was a sticky fog of occultists, wandering bishops, American intelligence and Nazi scientists. All were a handshake or two away from JFK's alleged assassin, Lee Harvey Oswald. They were all talking to ghosts, practicing ritual magic, holding hands around the séance table or sacrificing chickens in New Orleans apartments. And, in some cases, they were also members of America's ruling elite, the wealthiest and best-connected families in the country."[1]

The CIA was not the first organization to confront the task of probing the human mind. Operations Bluebird, Artichoke and MK-ULTRA were designed to respond to similar practices of the Soviet Union and China, what was termed in the popular press "brainwashing." The show trials of Cardinal Mindszenty of Hungary was more evidence that the Communists had managed to develop a technique for altering consciousness in their political prisoners; American GIs returning from captivity in Korea were another example. That there might be a mysterious Oriental method to "cloud men's minds" both frightened and excited the men of the CIA.

In order quickly to learn as much as possible, they assigned psychiatrists, scientists and medical professionals to the task of finding out how the mind, and specifically memory, works. In doing so, they came across occult practices, demonstrations of psychic abilities, and the mind control techniques of yogis, shamans and witch doctors. Clearly, this was a realm of mind control that was well worth pursuing, and which gave birth to some of the strangest projects ever funded by the US Government. It was in combination with the work of Lewin, Trist and Rees at Tavistock that these drugs, spiritual techniques and laboratory tests opened a Pandora's Box of suffering, violence and perhaps even redemption, through transformation: the Black Box of consciousness. In this, Tavistock, the world's premier brainwashing institute, was unwittingly following in the steps of magicians, sorcerers, gurus and cultists the world over. However, there was one difference.

Without meaning, there is no context for the experience; there is no way to integrate the material into one's psychological makeup. Tavistock's director, Brigadier General John Rawlings Rees wasn't interested in meaning, neither was Lewin or Trist or Adorno. That wasn't their job. Their task was to open up the mind to fast and easy manipulation, not to promote spiritual or psychic integration, or what Jung calls "individuation." Their job was to create assassins, turn agents, interrogate prisoners, obtain information, and manipulate consciousness. Saving souls was for the priests.

The CIA and the satanic cults; the mythology of the late twentieth century is surprisingly coherent even though the masks change from case to case, from victim to alleged victim. The CIA, of course, does exist; their mind control programs, from BlueBird to Artichoke to MK-ULTRA are a matter of public record. Their history of political assassinations and the overthrow of various foreign governments are also a matter of record. Satanic cults, or perhaps we should qualify that and say "occult secret societies," also exist and are a matter of public record; their attempts to contact higher spirits through arcane rituals are also well-documented and well-known.

Add all of this to espionage activities and you have a perfect fruit salad of paranoia, power and prestige, mixing secret government work with secret rituals and the manipulation of those sinister forces we have been looking at. The secret handshake of the cult and the secret code of the intelligence agency: they both evoke the power of relationship that is beyond the reach of ordinary human beings.

"Intelligence officers and cultists have a lot in common. Secrecy is a way of life for both the spy and the sourcerer; they both use codes and code names; they both pretend to have access to mysteries not available to the general public; they both claim to be able to influence events at a distance with their special abilities and power. They both specialize in the manipulation of reality; both are aware that things are not always what they seem to be; and they are both ruthless and often amoral or immoral in the pursuit of their goals. And when one can so easily manipulate the perception of reality, one eventually comes to the realization that Truth, itself, is a malleable thing. So it was only natural that the cultist and the spy would gravitate towards each other and would try to learn from each other."[2]

This control of reality, of the perception of reality and the creation of 'consensus reality' is a powerful political tool, and has been since ancient times when the proverbial sorcerers could appear to create solar eclipses by simply knowing when one would occur and act accordingly. To control and manipulate the reality of the masses, one uses what is known as psychological warfare. And it is but one step from controlling a person's perception of reality to getting that person to act on it.

The CIA and Tavistock were opening a Pandora's Box of demonic forces: the black box of consciousness. The techniques included drugs, various forms of hypnosis and even more extreme measures such as those developed by Dr. Ewen Cameron in Montreal, procedures known as "psychic driving," which involves drastic sensory deprivation sessions in an effort to wipe the consciousness clean and record a new consciousness over the old, much the same way we would record over a used cassette tape. It is the story of modern day Frankenstein, or a laboratory full of Frankensteins, and the monsters they made: monsters that wander the streets of our cities today. Quoting Peter Levenda: "The complexity of the human experience is such that we can only wonder what trigger mechanisms exist in environment – on television, in newspapers and magazines, and even the Internet – that suggest modes of behavior to these victims that are dangerous to themselves and to us.

"It would be only a matter of time before the mind control researchers began to scour the records of occultists, magicians, witches, voodoo priests and Siberian shamans to isolate the techniques that were used since time immemorial to supplant a person's normal, comfortable, everyday consciousness and replace it with a powerful, all-knowing and sometimes violent, and always deceptive alter personality; and to use those alters to uncover the action of deep memory, for MK-ULTRA was, at its core, an assault on the Land of Memory; the creation of new, false memories and the eradication of old, dangerous ones.

"For intelligence purposes, in the search for the perfection, what was required of MK-ULTRA was the ability to manipulate memory. But in exploring the mind and developing techniques for unlocking its secrets, the Agency unknowingly tread on areas that have been the domain of religion and mysticism for thousands of years. When the CIA incorporated drugs into the program and bizarre sensory deprivation techniques, we had all the elements of a serious occult experience. Cults as disparate as that of Eleusis in Greece, Tantra in India, Siberian shamanism, Native American shamanism, Taoism in China, Jewish Qabalism, even the relatively modern phenomenon of European ceremonial magic – as represented in the twentieth century by the Golden Dawn, the OTO, and individuals such as Aleister Crowley. As mythologists such as Carl Jung, Mircea Eliade and others have demonstrated, there is a great deal of similarity in 'technology' among these arcane practices, and that similarity exists for a reason."[3]

The arcane doctrine and methods of a discarded science are at the heart of this study, because they reveal the mechanisms by which society in general, and individuals in particular, have been manipulated by forces beyond their comprehension.

Again, quoting Peter Levenda's *Sinister Forces*: "Magic was there when MK-ULTRA began its search for the secrets of the paranormal and interviewed witches and wizards in America and beyond. Ceremonial magic begins with the basic premise that is sometimes formulated as the Hermetic axiom, 'As above, so below,' a simple phrase with ominous implications. Magicians believe that connections or links exist between perceived phenomena, and that to operate on one side of the link is to cause change to occur on the other. The magicians operate in the world of 'non-locality,' a world where the force may be an object, a wave may be a particle, and everything is in the immediate communication with everything else.

"The magician pulls his strings, surrounded by an aura of deniability, an aura created by Newtonian science, since science states that what the magician does is impossible, the result of superstition and ignorance. 'There is no such thing as the Mafia,' stated J. Edgar Hoover, Director of the FBI. And the capos laughed."[4]

Mind control and magic: not very far apart at all. "Thus, the magician is at once an interrogator and interogatee; he is the one in charge of manipulating and controlling the environment, but not for the effect it will have on someone else, but on himself alone. What the CIA has done in its interrogation manual – itself the product of MK-ULTRA – is to separate the magician from the ultimate goal of all occultism, which is the kind of spiritual perfection and elevated consciousness, and instead focus all the powers on occult techniques on an unwilling and uninformed subject, to manipulate him as well as the environment, to change the subject and transform him into something more useful to the interrogators and of mortal danger to the subject's own people. It is, in the jargon of occultism, black magic; and black magic in the service of the State."[5]

The doors of perception had been opened, not only by drugs like mescaline, LSD and psilocybin, but also by the séance, the shamans and the secret ritual. All had become tools of the trade for elements of the CIA, and their assault on memory and consciousness threw open the doors of perception and, instead of letting the Light in, they let Darkness out. Drugs, shamanism, and the occult. The dark domain of Charles Manson, of John Rawlings Rees, of Nazi doctors, of Hollywood and the music industry, the CIA and MI6 initiation sessions. With psychedelics, they disturbed the sleep of ancient forces, and the world would never be the same again.

ELIMINATION BY ILLUMINATION

Psychological warfare, although in existence for centuries, was really a "discovery" of the Second World War. Korea, the Philippines, Vietnam, Africa, the Middle East, Latin America, the roll call of psychological operations is long and, for the most part, classified. Taken individually, these psychological warfare operations may have had a succinct political purpose, as defined and identified by nameless men in grey flannel suits in the corridors of power, or by Presidents and National Security Advisors to further their own, hidden agendas; but they were manifestations of something deeper, a spiritual war, a war of world views, a war we are still fighting, numbly, into the second decade of the twenty-first century.

By 1964, the use of occult themes and rituals became an accepted part of psychological warfare planning. The Special Operations Research Office at American University "prepared a paper at the request of the U.S. Army on 'Witchcraft, Sorcery, Magic and Other Psychological Phenomena and their Implications on Military and Paramilitary Operations in the Congo.' The paper was authored by James R. Price and Paul Jureidini. American University was no stranger to psychological warfare research, having seen the establishment there of the Bureau of Social Science Research in 1950. Its African studies of psychological warfare were underwritten by the Human Ecology Fund, an organization that was a front for the CIA's MK-ULTRA program."[6] Some of the people involved in the research were the same Nazis who years earlier escaped to the United States as part of Operation Paperclip. The relationship between the occult and the Nazis shouldn't surprise anyone who understands that the highest ranking Nazis were contaminated with the ersatz paganism and popular occultism of the Thule Gessellschaft variety which had been absorbed into the SS with torchlight processions and runic chants.

OPERATION PAPERCLIP

As Levenda writes in *Sinister Forces*, "There had to have been a level at which the US government's leaders identified with the Nazis, or at least admired them. There had to have been a point at which the crimes of the Holocaust were considered a minor distraction, a kind of public relations nuisance, that was overshadowed by the glamour of the perfectly-run superstate of the Third Reich. There must have been an understanding that the ideologies of America and Nazi Germany were more alike than the ideologies of America and the Soviet Union. This is because there is just no other way to interpret what took place at the end of the war; morally, what transpired can only be considered a war crime itself.

"Policy makers in Washington knew that the next major conflict would be fought between the United States and the Soviet Union. It was of utmost importance that top level German scientists working on super secret V-1 and V-2 projects as well as nuclear weapons technology, be brought to American shores, out of reach of the Russians and, more importantly perhaps, pressed into service to the United States.

"To that end, several wartime intelligence operations were put in motion. The most famous one was Operation Paperclip. Most people who have heard of Paperclip assume that it was a program to bring Nazi scientists to the United States to assist the space program. However, there was much more to Paperclip, and to subsequent Nazi recruitment, than rocket science. Nazi medical personnel were also recruited, as well as psychological warfare experts and, with the Gehlen Organization, spies, assassins and saboteurs.

"The Paperclip story is long and very complex; it involves the alphabet soup of intelligence agencies and programs, from CROWCASS (Central Registry of War Crimes and Security Suspects) to CIC, SIS, OSS, CIA JIOA, and many more. It deals with dozens of countries, their intelligence agencies, armies, political parties, Roman Catholic Church, and criminal justice systems. By the time Paperclip was over, an entire division of Ukranian Waffen SS as well as thousands of Nazi scientists, many of them accused war criminals who had participated in some of the war's worst atrocities, had managed to find home in the United States, South America and in the Middle East."[7]

Whatever the morality of the recruiting of German scientists, there was another dimension to the experiments that has gone virtually unnoticed by researchers due to the near total lack of documentation and the lack of living witnesses. The evidence that does exist is largely circumstantial, but much of it can be found in the Captured German Documents section of the National Archives and then collated with later testimony that turns up in memoirs, biographies of the war years and its aftermath.

The first evidence we have that anything like a mind control program existed within the Third Reich is the memoirs of Himmler's astrologer, Wilhelm Wulff, "who talks about the Nazi desire to create a program within the Reich to duplicate the mental state of the Japanese soldier, a human being willing and eager to risk his life without question for his country, and the Chinese Communist 'human wave' trooper, who would rush unthinkingly into certain slaughter. Thus, Paperclip scientists were involved with military and the CIA mind control programs, such as Friedrich Hoffmann, a Nazi chemist who advised the CIA on matters relating to psychotropic substances for use in brainwashing."[8]

The Germans were among the first to study the use of psychological warfare; at the same time, after the war, psychological warfare became inextricably intertwined with propaganda and communications, and eventually bled over into acts that can only be considered terrorism: assassinations, sabotage, torture and interrogations, the domain of Tavistock Institute of Human Relations. As psychological warfare became more sophisticated, and the intelligence services at once more creative and more demanding, new techniques were developed and virtually codified.

All of these techniques share an ontological purpose: to manipulate perceptions and to re-create reality. Once that Pandora's box was open, there was no closing it again. The temptation was too great. For those who wanted to play God, there was the next best thing: one could play with the elements of creation in such a way that magical transformations would take place. As the men of the OSS, CIA, military intelligence and with Tavistock's oversight developed from the armchair scholars that most of them were before the war years into soldiers fighting on all fronts of the Cold War, they became, in a very real sense, magicians. "The CIA mind control projects themselves represented an assault on consciousness and reality that has not been seen in history since the age of the philosopher-kings and their court alchemists."[9]

MIND CONTROL

In an extraordinary book, *Secret Societies and Psychological Warfare*, Michael Hoffman writes, "looking into the subject of mind-control, one finds that the scope is wide and methods used are sophisticated. Mind control traces its origins to religious institutional use by priesthoods. Techniques of mind control developed in our western culture were field-tested by the Jesuits, certain Vatican groups, and various mystery religions, secret societies and Masonic organizations. Methods tested during the Inquisition were refined by Dr. Josef Mengele during the reign of the Third Reich.

"Soon afterwards, a mind control project called Marionette Programming imported from Nazi Germany was revived under the new name, 'Project Monarch.' The basic component of the Monarch program is the sophisticated manipulation of the mind, using extreme trauma to induce Multiple Personality Disorder (MPD), now known as Disassociative Disorder. In public testimony submitted to the President's Committee on Radiation, there are amazing allegations of severe torture and inhumane programs foisted upon Americans and other citizens, especially as children."[10]

This wedding of the purely psychological with the purely physiological became the cornerstone of subsequent intelligence agency programs designed to uncover the secrets of the mind: the interface between the lump of grey matter we call a brain and the great beyond we call reality.

THE CIA CONSPIRACY

As Peter Cuskie writes in *The Shaping of the Anglo-American SS by War*, "according to the OSS's Assessment Staff's post war report Assessment of Men, it was Dr. John Rawlings Rees and his Tavistock crew at the British War Office Selection Board, who contacted the London office of the OSS to suggest that the special warfare organization adopt Tavistock's selection and training methods. It was Rees who devised the OSS brainwashing 'election' procedures and Lewin helped refine them. It was also Rockefeller-sponsored Rees who devised an U.S. financier-promoted insurgency-counter-insurgency project. The development of the CIA establishment has been essentially directed to a systematic infiltration of all principal existing institutions, an infiltration, which is a deliberate, preparatory deployment, aimed at the establishment of a 'quasi-legal' fascist takeover with the help of some of the leading families of America."[11]

Many of the future leaders of the CIA came from America's ruling families, from a never-ending pool of bankers and industrialists, such as DuPont, Vanderbilt, Bruce, Mellon, Archbold, Morgan and Roosevelt. For example, Teddy Roosevelt's grandson Quentin Roosevelt was an OSS Special Operations officer in China as was Winston Churchill's cousin Raymond Guest. J.P. Morgan's two sons Junius and Henry S., "were in charge of the laundering of all OSS funds and the counterfeiting of all OSS identity papers."[12]

THE ROCKEFELLER CONNECTIONS

One of the people closely associated with the Rockefeller clan was John Gardner, an OSS psychologist in charge of "personnel assessment." For over three decades, he headed Rockefellers' most important organizations: The Rockefeller Brothers' Fund, the Carnegie Corporation, Education and Welfare Department and Common Cause, "a nationwide, independent, non-partisan organization for those Americans who want to help in the rebuilding of the nation."[13] In fact, Common Cause was a front for a cabal to impeach and remove Nixon from the White House on the one hand and install David's brother Nelson Rockefeller as the President on the other.

Arthur Goldberg, another one of Rockefeller's loyal servants, future U.S. Secretary of Labor and Supreme Court Justice under President John F. Kennedy, was placed in charge of the Office of Strategic Services (OSS) Labor Desk. Along with Trist and Lewin, Goldberg would push counterinsurgency as a method of labor movement control.

The list of Rockefeller supporters and representatives in key positions is a who-is-who amongst the "nation builders" of the XX century. Allen Dulles, former director of Rockefeller's Standard Oil Corporation, was OSS chief of Secret Intelligence for Europe and later head of the CIA. Dulles, a prototypical Eastern Establishment figure, is a perfect example of the invisible confluence of fascist interests around the creation of new imperial dominion controlled by Rockefeller through CIA-Tavistock corporate interests. For instance, Dulles was put in charge of the CIA's Bluebird mind control project, which he changed to Artichoke because he was fond of the vegetable. Furthermore, Allen and his brother John Foster were senior partners at Rockefeller's Standard Oil's chief law firm, Sullivan and Cromwell, a notorious CIA front with links to the most important financial houses on Wall Street.

Sullivan and Cromwell's "twin" was a German law firm of Albert and Westrick, who was simultaneously Hitler's financial agent, "an Abwehr spymaster in the United States and Sullivan and Cromwell's German representative. As a result, the Dulles law firm acquired three major German concerns as Standard Oil cartel partners. Among them was the murderous I.G.Farbenindustrie, which was, with the Kruppwerks, the principal user of concentration camp slave labor for the Nazis' blitzkrieg economy."[14]

The Dulles family had long and extensive ties to the Nazi underground, and had been laundering Nazi money for decades. "Both played major political roles throughout the 1920s and the 1930s in the 'Rearming of Germany by Night' policy, premised on using the Wehrmacht as a weapon against the Soviet Union and European working class."[15] Allen Dulles was in an intelligence position in Switzerland during the First World War, developing the contacts and the network he would use as the years went by. He would be one of the key people working the Byzantine machinations of the post-war Operation Paperclip that formed the real structure of resurgent fascism, of what became known in the intelligence community simply as Blowback.[16]

While Allen was employed in Switzerland, his brother John Foster was working at the State Department in Washington. Their uncle, ultra-conservative Robert Lansing, was President Woodrow Wilson's Secretary of State. The Dulles brothers actively courted Nazi officials in the 1930s, as did the Rockefeller and Wall Street corporate interests, such as Ford Motor Company and IBM and helped them escape after the war, especially when Allen Dulles became CIA Director beginning in 1953. Dulles' Chief of Counter Intelligence, James Jesus Angleton, later the CIA's Vatican liaison, and Frank Wisner who oversaw Gehlen Organization's 'Special Forces' in 1952-1953, would be involved with various secret off-the-books operations, among them Odessa, which helped relocate Nazis into Peron's Argentina and Die Spinne or The Spider, a loose designation of the post-war remnants of the Nazi secret police and covert intelligence apparatus, whose US point man was a Rockefeller Republican Party supporter, Harold Keith Thompson. We will come back to Die Spinne a bit later in the report.

In 1945, Dulles, along with former Gestapo OSS agent Hans Bernd Gisevius was accused by the US Treasury Department of laundering Nazi funds from Hungary into Switzerland. The investigation was eventually dropped when the US State Department claimed jurisdiction.

If anyone still has any doubts about a close proximity of the Rockefeller clan to the Nazis, in 1945, President Truman fired Nelson Rockefeller for his role in engineering the creation of Peron's fascist regime, which provided a 'safe haven' for over 100,000 Nuremberg criminals delivered by Allen Dulles from Europe.

In his national bestseller, *Blowback*, author Christopher Simpson wrote: "after World War II, Nazi émigrés were given CIA subsidies to build a far-right-wing power base in the U.S. These Nazis assumed prominent positions in the Republican Party's 'ethnic outreach committees.' These Nazis did not come to America as individuals but as part of organized groups with fascist political agendas. The Nazi agenda did not die along with Adolf Hitler. It moved to America (or a part of it did) and joined the far right of the Republican Party."[17]

From Richard Helms, a future CIA director and Dulles liaison man to Wehrmacht intelligence chief General Reinhard Gehlen to CIA director William Colby who received his training as the OSS Special Operations officer from the SOE in London, the list of interlocking Rockefeller-Nazi-Tavistock interests is endless. The Tavistock techniques of gang, counter-gang, sabotage and political assassinations Colby "learned at SOE's special training school were the very methods he later used in directing the CIA's Project Phoenix in South Vietnam"[18] which included the mass assassination atrocities at My Lai.

It is one thing to generate a conspiracy during a period of aggravated wartime crisis. It is another matter altogether to sustain and extend its dominion throughout the period of apparent peacetime normalcy. As things settled down at the end of the World War II, the Rockefeller-CIA set about to expand its apparatus. In the process, they would try to permeate and dissolve the old order "to emerge as the hard core of a new imperial dominion."[19] The Tavistock – and the Reesian method – that the class war should be waged with "weapons that affect morale more than they take life" – has become, in the post-war period, the primary weapons system of the Rockefeller forces, including their own covert arm, the Central Intelligence Agency.

For example, in 1977, the U.S. Senate Church Committee investigation exposed criminal activities by the Central Intelligence Agency. Quoting Peter Cuskie, "among the most damning revelations was the CIA's 25-year history of secret experimentation with mind-altering drugs, mass psychological manipulation, North Korean-style brainwashing and torture techniques, under the rubric of Operation MK-ULTRA, and other programs. The *New York Times* ran a front-page story, singling out Dr. Louis West as one of the CIA's top MK-ULTRA mind-destroyers."[20] The *Times* obtained a West-authored memo advocating the use of LSD in social control. The memo, the paper reported, read in part, "This method, foreseen by Aldous Huxley in *Brave New World* (1932), has the governing element employing drugs selectively to manipulate the governed in various ways. In fact, it may be more convenient and perhaps even more economical to keep the growing number of chronic drug users (especially of the hallucinogens) fairly isolated and also out of the labor market, with its millions of unemployed. To society, the communards with their hallucinogenic drugs are probably less bothersome – and less expensive – if they are living apart, than if they are engaging in alternative modes of expressing their alienation, such as active, organized, vigorous political protest and dissent."[21]

Earlier, that is shortly after the war, Rees gave a speech in New York to a group of military and civilian psychiatrists. "If we prepare to come out into the open and to attack the social and national problems of our day, then we must have the shock troops and these cannot be provided by psychiatry based wholly in institutions. We must have mobile teams of psychiatrists who are free to move around and make contacts with the local area."[22] The overall theme of all of Rees's known work is the development of the use of psychiatry as a weapon of the ruling class. Rees' 22-point program for the "application of military methods to civilian life," presented in his lectures that constituted the book *The Shaping of Psychiatry by War*, became an Anglo-American SS Bible.[23] From Vienna in 1956, Rees, under the cover of his World Federation for Mental Health (WFMH) and Frank Wisner, the CIA Deputy Director and chief of Special Operations coordinated the counterinsurgency insurgency popularly known as the "Hungarian Revolution."

Rees emphasized the value of overt and covert operations through the use of military as a model of improvements in civil affairs institutions. One of these covert operations was the Die Spinee network.

Through the period of the Cold War, the Die Spinne network was employed in a covert corporate and CIA-intelligence spiritual crusade to roll back the Iron Curtain, such as the paramilitary operation known as the Hungarian Uprising in 1956. After the 1963 JFK assassination, Die Spinne was de-activated in favor of the "Fascism With a Human Face" policy, championed by such left-fascist elements as the Institute for Policy Studies.

Executive Intelligence Review on Die Spinne operations makes an interesting observation, "of Cold war operations, the Hungarian Revolt of 1956 was Die Spinne's most famous. Amid a full deployment of British Fabian and similar operatives to destabilise Eastern Europe's Communist Parties following Khrushchev's 'De-Stalinization' speech, the Gehlen Organization's special forces comprised from former Waffen SS and Brandenburg Division Cadre, covertly invaded Hungary and linked up with old fascists from the Arrow Cross Party which the Nazis used during the occupation, to conduct a wave of sabotage and assassinations. It was these elements which formed the core of the so-called 'Hungarian Freedom Fighters.' The command structure extended down through scenario-design and policy-direction at the levels of National Security Council, the Foreign Intelligence Advisory Board, the Hudson Institute and the Hoover Institute of War."[24]

In the end, the CIA-directed liberation of Hungary failed because when push came to shove, the Soviet Union was willing to undertake a nuclear war to preserve their hold over the satellite states. The Americans, though rhetorically committed to liberation, were not willing to fight World War III to achieve that objective. "Among the Nazis laundered through Allen Dulles' Operation Sunrise and Operation Paperclip, for post-war use were SS Untersturmfuhrer Paul Dickopf who would serve as Interpol's head between 1968 and 1972. Interpol is an international private police agency which functions as a coordinating body for Die Spinne terrorism."[25]

A little known fact about Interpol is that it grew out of anticipation of the First World War, set up by the House of Rothschild in Vienna in 1923. The family felt it needed a special intelligence organization to watch over the interests of bankers, who were financing both sides of the war. In order not to make things look too suspicious, they called on Prince Albert I of Monaco to invite lawyers, judges and police officers from several countries to discuss international co-operation against crime.

With the abundance of proof, still most people would find it difficult to accept that the CIA is an adjunct of the Rockefeller-led organization that along with Tavistock Institute is working diligently to change the paradigm of modern society. If we are to believe the United States Federal statutes and other myths concocted for the benefit of the credulous, the CIA is merely a key agency of the National Security Council, working with various intelligence agencies, the Defense Department, the Treasury and the FBI. The reality is quite different. The CIA apparatus has penetrated every key organization, including think tanks, universities, non-governmental organizations and foundations.

For example, most major universities are either entirely or substantially branches of the CIA. At the University of Michigan, we have the most notorious example of the Institute for Social Research (ISR). Calling its work a "telescope on society,"[26] ISR are notorious disciples of John Rees and Tavistock. Then, there is Harvard's 'Russian' institute, like Columbia's, but also Harvard's psychology department is a covert CIA operation run for many years by another Reesian psychopath, B.F. Skinner, a psychologist who asserted that men have no minds. Massachusetts Institute of Technology has always been a flourishing nest of CIA activity. The University of Pennsylvania's Wharton School was home of Eric Trist, a Rockefeller-sponsored Reesian fascist, who directed a series of social work projects. Cornell, Berkeley and Stanford are also notorious CIA fronts.

Through control of governmental and major foundations, the CIA has been able to have a say "not only what programs are funded, but to control the selection of instructors who move into controlling positions"[27] as honest academicians are weeded out and hounded into retirement.

Then, there is the case of CIA's control of private foundations. For example, the Rockefeller Foundation was established by the family on April 24, 1913. The statement of purpose cynically read, "To promote the well-being of mankind throughout the world." Another powerful Rockefeller controlled foundation is Ford Foundation, whose head was a Rockefeller stooge and a CIA operative, McGeorge Bundy of the CIA Bundy family. McGeorge's father, Harvey, married a Lowell, one of the wealthiest American families and then worked under Henry Stimson at the War Department. After helping Stimson write his memoirs, Bundy joined Richard Bissell, his professor at "Yale during his undergraduate days, in Europe to help implement the Marshall Plan. When Bissell became head of Special Operations at the CIA, Bundy went with him again."[28] Bissell later would become a Ford Foundation director and a member of the Council on Foreign Relations. What's more, William Bundy, McGeorge's older brother was not only one of the most powerful men in the CIA, but was also a member of the Rockefeller-controlled Council on Foreign Relations and an editor of CFR's *Foreign Affairs* magazine.

The Rockefellers also have interlocking control over the other most powerful foundation, The Carnegie Foundation. Leading directors of the Carnegie foundations have for decades belonged to the Rockefeller-coordinated, Council on Foreign Relations. In addition, twelve of the seventeen members on the Carnegie's Board of Trustees are also associates of various Rockefeller financial institutions, including Richard Beattie, chairman of Simpson Thacher & Bartlett LLP law firm, a notorious CIA front and Susan Hockfield, President of Massachusetts Institute of Technology. Another Rockefeller supporter is a current president of Carnegie Endowment for International Peace, Jessica Tuchman Mathews who is also a regular member of the powerful Bilderberg organization. Carnegie is a foreign police think-tank interlocked with other Rockefeller think-tanks and foundations such as Ford, Hoover Institute and Hudson Institute, whose senior fellow, Marie-Josée Kravis is also a member of the Bilderberg group.

When I think of this chapter, I think of Mona Lisa, or rather the smile of Mona Lisa. Her grin is the grin of men and a few women who engage in the most vicious forms of psychological warfare against the human race. The knowing grin of men who commit crimes against humanity. In this chapter, we have exhumed the corpse of Tavistock's greatest architect, John Rawlings Rees and a few of his most loyal henchmen, tracing the living trail of mind control and ritual killing and placing them at the feet of some of our most important puppet masters.

Let there be Light, He said. The skeleton key turned once, twice, three times. The door was pulled ajar with a squeak. Something moved. A shadow. Then, it was gone.

*****

There is blood on our hands, and there are documents. This is history. You can't have one without the other. Blood. Documents. Guilt. Innocence. Knowledge. Ignorance. Frustration. Fear. But you can't know history unless you know fear. You can't know history unless you feel the pulse of life under your fingers; unless you can stare the guns in the face. Unless you can stand in the prisons, and the death camps, and feel the gaze of informers and spies and soldiers on your back in foreign countries … and on your own doorstep, your own driveway. History is not to be absent but to become absent; to be someone and then go away, leaving traces. The rest is only bookkeeping.

The world has always been like this, of course. It has always been run by people: superstitious, religious, fearful, paranoid, ugly, hateful, murderous people. That is nothing new. But at a specific point in the century we took this a step further. "With JFK, we opened Pandora's box, and the black box of human consciousness. We flipped open the lid, and rummaged around inside. And we loosed monsters on the earth. Monsters who feed on human flesh, and who drink the nectar of human souls.

"With what we have seen here, one comes to the conclusion that there is nowhere to go, no one to trust. Everything we know, we see through a prism of a lens, cold, and calculating. For what else is a camera but a special device for transforming and viewing our world? As any photographer who has slipped beneath the dark cloth and focused a portion of the world upon their ground glass can tell you, the eloquent mirror of the lens inverts and reverses the world – optically, mathematically and magically. As we pose our subjects and view them through our lens we create our own special world and, emerging from the dark cloth are an array of characters who made the XX century their own – JFK, Rockefeller, Rees, the little grey men in flannel suits and CIA badges, their Tavistock companions and all sorts of unsavory, crazy Nazis. It is The Wizard of Oz meets MK-ULTRA. We live in the most at the abnormal – even paranormal – of times. And unless we heed the reality that underlies the everyday, our failure to connect the dots will allow our denial to persist. With JFK front and center, we are awash not in the coincidental but in the subliminal, and the possibility that the mythic battle between demonic and angelic forces are yet being recreated on planet Earth. We could end this by saying, "We are not in Kansas, anymore." Unfortunately, might I say, paraphrasing Mephistopheles, 'Why, this is Kansas. Nor am I out of it.'"[29]

Endnotes

1. *Sinister Forces – The Nine, Book 1*, Peter Levenda, Trineday Press, May 2011

2. Ibid

3. *Sinister Forces – The Nine, Book 1*, Peter Levenda, Trineday Press, p.219, May 2011

4. Ibid

5. Ibid

6. Christopher Simpson, *Science of Coercion: Communication Research & Psychological warfare 1945-1960*, Oxford University Press, NY 1994, ISBN 0-19-510292-4

7. *Sinister Forces – The Nine, Book 1*, Peter Levenda, Trineday Press, p.134-135, May 2011

8. Ibid

9. Ibid

10. *Secret Societies and Psychological Warfare*, Michael Hoffman, Independent History and research, 2001

11. Peter Cuskie, The Tavistock Grin, The Shaping of the Anglo-American SS by War, *The Campaigner*, May 1974

12. Ibid

13. http://www.commoncause.org/site/pp.asp?c=dkLNK1MQIwG&b=4860209

14. Die Spinne: How Rockefeller kept the Third Reich Alive, *EIR*, Volume 3, #45, November 8, 1976

15. Die Spinne: How Rockefeller kept the Third Reich Alive, *EIR*, Volume 3, #45, November 8, 1976

16. Christopher Simpson, *Blowback*, Weidenfeld & Nicolson, 1988

17. Ibid

18. Peter Cuskie, The Tavistock Grin, The Shaping of the Anglo-American SS by War, *The Campaigner*, May 1974

19. Ibid

20. Ibid

21. Ibid

22. The Tavistock Grin, *The Campaigner* magazine, May 1974, p45

23. Ibid

24. Die Spinne: How Rockefeller kept the Third Reich Alive, *EIR*, Volume 3, #45, November 8, 1976

25. Die Spinne: How Rockefeller kept the Third Reich Alive, *EIR*, Volume 3, #45, November 8, 1976

26. http://www.isr.umich.edu/home/about/

27. Rockefeller's Fascism with a Democratic Face, *The Campaigner*, Vol. 8, #1-2, November-December 1974

28. Peter Cuskie, The Tavistock Grin, The Shaping of the Anglo-American SS by War, *The Campaigner*, May 1974

29. *Sinister Forces – The Nine, Book 1*, Peter Levenda, Trineday Press, May 2011

CHAPTER THREE

Killing of the King

E veryone loves listening to music, but how many really understand the meaning of the lyrics and its effects on the listeners. Many of the mainstream songs you hear that make it to the top of the Billboard charts have very important meanings and significance. We tend to ignore things we don't understand, focusing our attention on the words themselves instead of the deeper meaning behind the words such as imagery, symbolism, historical references and the Occult within the lyrics. Combined with visual representation of the songs, our senses are bombarded by a whole range of emotions that appeal to our subconscious.

What I am about to tell you may shock you so profoundly that you think twice about reading this. Be forewarned. You don't have to believe it, but if you are prepared to disbelieve it, at least give me the benefit of reading it through, once.

WORDS WE USE

M ost people never remember the whole song but snippets of a song, mostly the chorus that consists of catchy words or play on words. We hum the song along with the radio until we come to the part of the chorus where for a few brief lines we happily sing along with the artist, mindlessly repeating the words we memorized. Let's take a look at some songs. You will start to notice a pattern.

We are in a crowded auditorium. Lights go out and the camera zooms in on two individuals on stage. A white man with a red baseball hat and a fouled-mouthed black man in a white beret. "What the fuck are you talking about?" shouts the black man. The white man, slowly walks across the stage, then pauses. "If I do come back, do you know what my new name is gonna be?" he spits back to his friend. "What's your name gonna be?" asks the man in the white beret with an in-your-face attitude. "Rain Man," replies the white rapper, whose stage name is Eminem. There is a heavy drum roll, accompanied by piped in music from somewhere above. "Put your hands in the air," shouts Eminem, gesticulating wildly at the audience gone numb from the noise and the excitement of the moment. "Say, Rain Man," shouts Enimen to the crowd. "Rain Man!" more than 50,000 people at Yankee Stadium in New York shout back, hands in the air, waving back and forth. Over the next 4 minutes, the words "Rain Man" are repeated over 40 times by the performers on stage and an equal number of times by the frenzied crowd. In between, the two men sing in unison

> *You find me offensive,*
> *I find you offensive for finding me offensive*
> *I aim to be calm when I got to this level of fame*
> *Possessed by this Devil, by the name...*
> *Rain Man*

Later, in Eminem's "Old Time's Sake," he returns to the familiar theme.

> *Speak of the Devil*
> *It's attack of the Rain Man.*
> *Chainsaw in hand.*
> *Blood stain on my apron.*

In fact, Eminem, is one of a large number of mainstream and highly popular performers who sing about a someone called Rain Man. Others who use the same motif in their performances are Rihanna in her song "Umbrella," where the words Rain and Umbrella, appear in the chorus and throughout the song, accompanied by images of Rihanna and her dance troupe soaking wet with rain falling on their heads. Others, such as rapper Fat Joe Feat and his "Make it Rain," someone called Savage and his "Wild Out," Black Stone Cherry and their popular rendition of "Blind Man" and Jamie Foxx with his particular version of "Rain Man." Is it a coincidence that so many mainstream artists sings about a someone called Rain Man? Or is there more to it? Let's take a closer look at the words themselves. Foxx sings the following in the chorus:

> *Release me.*
> *Ohhhhh uhh ohhhhhhh.*

Ohhhhh uhh ohhhhhhh
Rain Man, Rain Man, Rain Man

Rappers, The Game, in their song, "Cali Sunshine" sing:

With more ass then delicious.
That's my flower of love.
We make it rain like rain man.
When we play with the glove.

In Jay-Z Feat. Usher & Pharrell's song "Anything," we have

Amateur pole dancing.
Come and get this cash from me.
They call me Rain Man.
She tried to rain dance.

Rappers are not the only once to use the term Rain Man in their lyrics. Heavy metal band, W.A.S.P. in their song "The Burning Man," also mention it. In Bob Dylan's "Stuck Inside of Mobile with the Memphis Blues Again," that appears on his 1966 album *Blonde on Blonde,* he sings the following

Now the Rain Man gave me two cures.
Then he said jump right in

And again in "I Wanna Be Your Lover."

Well the Rain Man comes with his magic wand
and the judge says Mona can't have no bond

Hip Hop artist Ya Boy pays tribute to the Rain Man in his likewise titled track.

Hey everybody say Mr. Rain Man can we have a rainy day?
Bring an umbrella Please bring an umbrella ella ella ella ehh.

Former country music superstar Tanya Tucker also sings up to the Rain Man in a track titled "Lizzie and the Rain Man."

But there was one named Lizzy Cooper,
Who said he was a lying cheat.

She said you call yourself the Rain Man
Well you ought to be ashamed.

Then, things get really interesting and very explicit and in your face. For example, listen to Jay-Z's D´evils.

Dear God. I wonder can you save me?
Illuminati want my mind, soul and my body...
Dear God. I wonder can you save me?
Secret Society. Tryin´ to keep their eye on me...

And to Tupac who was killed in what is believed to have been an assassination.

Some say they expect Illuminati take my body to sleep.
Niggas at the party with their shotties just as rowdy as me.

A superstar rapper, Fatboy Slim also mentions them in his song titled, Illuminati.

Illuminati, a secret society do exist, Illuminati-ti-ti-ti-ti-ti.

And LLCool J in "I shot ya"

Illuminati want my mind and my body. Secret Society
Trying to keep their eye on me.

Now, instead of reading these lyrics as a form of amusement, ask yourselves: what if what they are singing is for real? What if someone is trying to tell you something in such an unbelievable, Tavistockian way, in your face, where you could best see it? What if the people who are trying to tell you something are the very secret societies and brainwashing institutes who are trying to control the world from behind the scenes? And what if I told you that there is no such thing as coincidences in the world of spies and secret societies, in the world of smoke and mirrors, in the parallel universe where governments work their secret agenda from behind the curtain? Now, read the next set of lyrics from Prodigy. The track is titled "Illuminati."

It took me 33 years for me to see the truth.
Ever so clear I
Was too young

I couldn't articulate myself the right way
Son, but now let me break it down, pass it all around, this
Is not a theory.
The conspiracy is real.
They wanna put me in
A straight jacket in a padded room and tell the world
There's 12 monkeys so they can be confused.
Illuminati want my mind soul and my body
Secret Society trying to keep their eye on me.

So, why would all these allegedly uncooth rappers, country musicians and rockers sing about a something called the Illuminati, the New World Order, Rain Man and Umbrella? In fact, it is becoming even more widespread and common. Listen to a rapper Puff Daddy in "Young G's":

Niggaz that got killed in the field and all the babies born
Know they ain't fully prepared for this New World Order.

And it's not just lyrics that contain these references. Music videos are also full of occult symbolism such as Ankh (ancient Egyptian symbol for Eternal Life), all-seeing Eye of God, Pentacles and pyramids, not to mention the Masonic black and white checkered floor. We can imagine what a New World Order is and we have heard of the Iluminati, thanks in part to Dan Brown's popular fiction, but who is this Rain Man?

WHY IN PLAIN SIGHT?

You might be wondering why, if a conspiracy of silence to control the world truly exists, the supposed Masters of the Universe, instead of hiding it, elect to promote it through songs, videos and Hollywood film. The truth of the matter is that secret societies love to hide things in plain sight. And not only through videos and lyrics. What if I told you that the JFK assassination was a ritual slaying, what is known in the world of the occult as "the killing of the King?" What if I gave you enough evidence to show you that this is so, despite the official version of the events, wherein we are told a lone assassin killed the President with a magic bullet?

To understand the Kennedy assassination and place it within the realm of Tavistock mind control, secret societies, One World conspiracy and Nazi occult, you must open the doors of perception into a nether world of the unknown by changing the way we look upon the details and the fantastic convergences of life. This is the way ancient man perceived the world, "encompassing a vision that detects every link and every symbol, beginning with the significance of names, then places, necessarily including lines of latitude and longitude and the divisions of degrees in geography and cartography and then the obsessive actions which stem from the confluence of the two and which have come to be known as ritual."[1]

Symbolism and other types of references can be flagrant and subconsciously picked up, but on a conscious level will not be seen by the naked eye. In film and videos, this type of subliminal messaging is achieved using scenes that contain brief flashes of light and dark in quickly alternating video frames.

At times, the symbolism is blatantly clear and easily identifiable, but because the meaning or references behind it are unknown to most people, the brain simply filters it out as insignificant and meaningless. However, none of it is meaningless. You will be shocked to see how nothing was left to chance in the JFK assassination, despite the apparent scenes of chaos and disconcert.

In penetrating the secrets of secret societies and symbols, it would be helpful to consider a dictum of Einsteinian physics, write James Shelby Downard and Michael A. Hoffman II: "Time relations among events are assumed to be first constituted by the specific physical relations obtaining between them."[2] Secret societies, Masonic lodges, cults and spooks use secret symbols to show their affiliation. The eternal pagan psychodrama is escalated under these 'modern' conditions precisely because sorcery is not what 20th century man can accept as real.

In their underground bestseller, "King-Kill/33," James Shelby Downard and Michael A. Hoffman II explain the mechanics of the assassination: "Keep in mind that the ultimate purpose of John Fitzgerald Kennedy's assassination was not political or economic but sorcerous: for the control of the dreaming mind and the marshalling of its forces is the omnipotent force in this entire scenario of lies, cruelty and degradation. Remember that once you spread the deadly virus, the disease will do the rest.

Something died in the American people on November 22, 1963 – call it idealism, innocence or the quest for moral excellence. It is the transformation of human beings, which is the authentic reason and motive for the Kennedy murder."[3]

The more one studies mind control, secret societies and the occult, "the more one understands the subtlety and hidden symbolism, the so-called 'science of names' word wizardry behind the scenes of Kennedy assassination. The JFK assassination encounters this science in a decisive way and contains a veritable nightmare of symbol-complexes having to do with violence, perversion, conspiracy, death and degradation."[4] Let's take a closer look.

THE KILLING OF THE KING

James Shelby Downard's underground classic *Sorcery, Sex, Assassination and the Science of Symbolism*, links American historical events with a grand occult plan of secret societies. His investigative work on the startling links within the Kennedy assassination has remained unmatched long after his death. Downward writes,

"President Kennedy and his wife left the Temple Houston and were met at midnight by tireless crowds present to cheer the virile 'Sun God' and his dazzlingly erotic wife, the 'Queen of Love and Beauty,' in Fort Worth. On the morning of November 22, they flew to Gate 28 at Love Field, Dallas, Texas. The number 28 is one of the correspondences of Solomon in kabbalistic numerology; the Solomonic name assigned to 28 is 'Beale.'

"On the 28th degree of latitude in the state of Texas is the site of what was once the giant 'Kennedy ranch.' On the 28th degree is also Cape Canaveral from which the moon flight was launched – made possible not only by the President's various feats but by his death as well, for the placing of the Freemasons on the moon could occur only after the Killing of the King. The 28th degree of Templarism is the 'King of the Sun' degree. The President and First Lady arrived in Air Force One, code-named 'Angel.'

"The motorcade proceeded from Love Field to Dealey Plaza. Dealey Plaza is the site of the Masonic temple in Dallas (now razed) and there is a marker attesting to this fact in the plaza.

"Important 'protective' strategy for Dealey Plaza was planned by the New Orleans CIA station whose headquarters were a Masonic temple building. Dallas, Texas is located ten miles north of the 33rd degree of latitude. The 33rd degree is the highest in Freemasonry and the founding lodge of the Scottish Rite in America was created in Charleston, South Carolina, exactly on the 33rd degree line.

"Dealey Plaza is close to the Trinity River, which before the introduction of flood control measures submerged the place regularly. Dealey Plaza therefore symbolises both the trident and its bearer, the water-god Neptune. To this trident-Neptune site came the 'Queen of Love and Beauty' and her spouse, the scapegoat, in the Killing of the King rite, the 'Ceannaideach' (Gaelic word for Ugly Head or Wounded Head). Now, get this: the "Queen" is Jackie and "Ceannaideach" is the Gaelic form of Kennedy. In Scotland, the Kennedy coat of arms and iconography is full of folklore. Their Plant Badge is an oak and their Crest has a dolphin on it. Now what could be more coincidental than for JFK to get shot in the head near the oak tree at Dealey Plaza. Do you call that a coincidence?

"At 12:22 p.m. the motorcade proceeded down Main Street toward the Triple Underpass, travelling first down Elm St. The latter was the scene of numerous gun fights, stabbings and other violence, and it is the location of the Majestic Theatre, the pawn shop/negro district, and industrial district.

"Elm, Main, and Commerce form a trident pattern in alignment with the triple underpass as any Dallas map will show. Many analysts contend that at least three assassins were involved in the crossfire ambush of Kennedy.

"It is a prime tenet of Masonry that its assassins come in threes. Masonic assassins are known in the code of the lodge as the 'unworthy craftsmen.' Hiram Abiff, architect of Solomon's Temple and mythic progenitor of Freemasonry was murdered according to Masonic legend by three 'unworthy craftsmen.' Because Masonry is obsessed with earth-as-gameboard (tessellation, thus the checkered Masonic floor) and the ancillary alignments necessary to facilitate the 'game,' it is inordinately concerned with railroads and railroad personnel to the extent that outside of lawyers and circus performers, no other vocation has a higher percentage of Masons than railroad workers.

"Minutes after John Fitzgerald Kennedy was murdered three 'hoboes' ('unworthy craftsmen') were arrested at the railyard behind Dealey Plaza.

"No records of their identities have ever been revealed nor the 'identity' of the arresting officer. All that remains of those few minutes are a series of photographs, a ritual accompaniment of the Black Mass that was the ceremonial immolation of a king, the unmistakable calling card of Masonic murder, the appearance of Jubela, Jubelo and Jubelum, the three 'unworthy craftsmen,' 'that will not be blamed for nothing.' This ritual symbolism is necessary for the accomplishment of the alchemical intention of the killing of the 'King of Camelot.'

"Dealey Plaza breaks down symbolically in this manner: "Dea" means "goddess" in Latin and "Ley" can pertain to the law or rule in the Spanish, or lines of preternatural geographic significance in the pre-Christian nature religions of the English.

"The systematic arrangement and pattern of symbolic things having to do with the killing of Kennedy indicates that he was a scapegoat in a sacrifice. The purpose of such macabre ritualism is further recognizable in patterns of symbolism culminating in the final 'making manifest all that is hidden.'

LEE HARVEY OSWALD

"**O**swald means 'divine strength.' The diminutive form of the word is 'os' or 'Oz': denoting strength. Divine strength is integral to the King-killing rite. The role which 'Divine Strength' played in the Dealey 'Goddess Rule' Killing of the King ritual should be given careful consideration.

"One should also note the significance of (Jack) Ruby's killing (destroying) of 'Ozwald' in reference to the 'Ruby Slippers' of The Wizard of Oz which one may deride as a fairy tale but which nevertheless symbolizes the immense power of 'ruby light,' otherwise known as the laser.

"Oswald may have undergone biotelemetry implantation in the Soviet Union while a 'volunteer' at a Behavior Control Center at Minsk. Oswald roomed with Cubans and was allegedly friendly with a Castro-man identified only as being a 'key man.' The 'key,' of course, is one of the most important symbols in Masonry and the symbol of silence.

ARLINGTON NECROLOGY

"The Kennedy and Oswald burials were both at 'Arlington':
JFK at the National Cemetery near Washington, D.C., and
Oswald at Rosehill Cemetery near Arlington, Texas. 'Arlington' is a
word of significance in Masonic sorcery and mysticism and it has
a hidden meaning that ties in with necrolatry.

"At the Kennedy gravesite there is a stone circle and in its middle
a fire that is called an 'eternal flame.' The fire in the middle of the
circle represents a point in the circle, the same type of symbolism
that is recognisable in Kennedy's bier and coffin being in the center
of the rotunda in the Capitol. A point in a circle symbolized the
sun in ancient sun worship. It was also a symbol of fecundity, with
the point symbolizing a phallus and the circle a vagina.

FUNERARY RITES

"In the ancient mysteries the aspirant could not claim a
participation in the highest secrets until he had been placed in
the Pastos, bed or coffin. The placing of him in the coffin was called
the symbolical death of the mysteries, and his deliverance was
termed a rising from the dead; the 'mind,' says an ancient writer
quoted by Stobaeus, is afflicted in death just as it is in the initiation
in the mysteries. And word answers to word, as well as thing to
thing; for burial is to die and death to be initiated. The coffin in
Masonry is found on the tracing boards of the early part of the last
century, and has always constituted a part of the symbolism of the
Third Degree, where the reference is precisely to the same as that
of the Pastos in the ancient mysteries.[5]

"President Kennedy sat at the head of a coffin table at the White
House. To his back, over a fireplace, hung a portrait of Abraham
Lincoln, an assassinated president. On either side of the picture
were urns that resembled the type called 'cinerary urns' which are
vessels in which the ashes of the dead are kept.

"A book about JFK was called *Three Steps to the White House*.
In Masonry are what is known as the 'three symbolical steps.' 'The
three grand steps symbolically lead from this life to the source of
all knowledge.'[6]

"It must be evident to every Master Mason without further
explanation, that the three steps are taken from the darkness to a

place of light, either figuratively or really over a coffin, the symbol of death, to teach symbolically that the passage from darkness and ignorance of this life through death to the light and knowledge of eternal life. And this from earliest times was the true symbolism of the step.[7]

"The body of President Kennedy was placed in a coffin, which was positioned in the center of a circle under the Capitol dome. The catafalque was "a temporary structure of wood appropriately decorated with funeral symbols and representing a tomb or cenotaph. It forms a part of the decorations of a 'Sorrow Lodge.'" This Masonic Encyclopedia entry refers to the ceremonies of the Third Degree in Lodges of the French Rite.

"Pictures taken of the Kennedy coffin and catafalque show these two props of the funerary rite as a point in a circle. Fecundity is the symbolic signification of the Point within a circle and is a derivation of ancient sun worship.

"In the lore of mystery cults and fertility religions was invariably the legend of the death of the hero god and the disappearance of his body. In the subsequent search and supposed finding of the body we see the contrivance of an elaborate psychological ruse. The body was said to have been concealed by the killer or killers of the hero god. The concealment of the body was called 'aphanism' and is a rite of the Masonic 3rd Degree. Anyone interested in comprehending the mechanics of group mind control would do well to study the 3rd and 9th degrees in particular, and all the grades of Masonry in general. The disappearance of the body, this aphanism, is to be found in the assassination of President Kennedy:

"The President's brain was removed and his body buried without it ... Dr. Cyril Wecht, chief medical examiner of Allegheny County, Pennsylvania, past president of the American Academy of Forensic Scientists, and a professor of pathology and law, received permission from the Kennedy family in 1972 to view the autopsy materials (at the National Archives) ... When he routinely asked to see the brain, Wecht was told it was missing, along with the microscopic slides of the brain. Marion Johnson, curator of the Warren Commission's material at the Archives said, "The brain's not here. We don't know what happened to it."[8]

"If and when the brain is recovered, the entire process will have been completed under the term 'euresis.' In the Masonic Mysteries are 'symbolical ladders'. On the Masonic tracing board of 1776 there is a ladder with three steps, a significant revision of the usual ladder in such references (seven steps).

"There are of course all sorts of ladders: the Brahmanical Ladder (seven steps), the Kadosh Ladder (seven steps), Rosicrucian Ladder (seven steps), Jacob's Ladder (various numbers attributed), the Kabbalistic Ladder (ten steps); then there is old "Tim Finnegan's Ladder" which is known to some as the 'Ladder of Misfortune', and it is seemingly comprised of one false step after another.

"After the Kennedy coffin was removed from the center of the Capitol rotunda circle, it was taken, with pageantry, to the street for viewing. The funeral procession made an 'unplanned stop' on Pennsylvania Avenue in front of the 'Occidental Restaurant' and a picture was taken of the flag draped Kennedy coffin with the word 'Occidental' featured prominently over it. In Masonry and in the lore of the Egyptian jackal-god Anubis, a dead person is said to have 'gone west'.

"Several months after the Kennedy funeral, 'Occidental Life', an insurance branch of the Transamerica Corporation, ran an advertisement for group life insurance which it proclaimed to be 'new' but contained a turn that was indeed original: the inferential weird claim was made that "Until now there was only one way to cash in on Group Insurance"; apparently some rather profound changes were made in the manner of things-as-they-are after the 'Killing of the King' had become a *fait accompli.*

"Before JFK began his *Jornada del Muerto* (Journey of Death), he was photographed with Yugoslavian dictator Tito on the winding stairs in the White House. Tito is a significant name in Masonry since it was the title given to Prince Harodim, the first Judge and Provost said to be appointed by King Solomon. The Tito was a reputed favorite of Solomon, whose temple was a hotbed of thievery, money-changing, male and female prostitution and sorcery. This ancient Tito presided over the Lodge of Intendants of this temple and was one of the 'twelve knights of the twelve tribes of Israel'.

"President Kennedy preceded Tito down the stairs to a portrait of the assassinated President Garfield where he was photographed, and another picture was taken on the stairs before a picture of Lincoln (the Lincoln Continental limousine in which Kennedy was shot).

"John F. Kennedy, the one and only Catholic president of the United States, was a human scapegoat, a 'pharmakos.' 'Pharmakos' or 'Pharmak-vos' can mean "enchantment with drugs and sorcery" or "beaten, crippled or immolated." In alchemy, the Killing of the King was symbolised by a crucified snake on a tau cross, a variant of the crucifixion of Jesus.

"Jesus Christ was tortured and murdered as the result of the intrigue of the men of the Temple of Solomon who hated and feared Him. They were steeped in Egyptian, Babylonian and Phoenician mysticism.

"Masonry does not believe in murdering a man in just any old way and in the JFK assassination it went to incredible lengths and took great risks in order to make this heinous act of theirs correspond to the ancient fertility oblation of the Killing of the King."

Michael A. Hoffman explains that the three hoboes arrested at the time of the assassination in Dallas were at least as important symbolically as operationally and that they comprise the 'Three Unworthy Craftsmen' of Masonry. "This symbolism is at once a telling psychological blow against the victim and his comrades, a sign of frustrated inquiry, the supposedly senseless nature of any quest into the authentic nature of the murderers, and a mirror or doppelganger of the three assassins who execute the actual murder.

"'Three Unworthy Craftsmen' of Masonry, part of the rites of initiation for the Master Mason, which is a third degree of masonry, are dramatic, as they "reveal the most enduring and most important mystery of all Masonic ritual: the legend of the murdered Master. After brief ceremonies, the candidate is blindfolded as the Worshipful Master begins to tell him the story of the murder of Hiram Abiff, the master builder of Solomon's temple. He explains that three assailants who demanded they be told the secrets of a Master Mason attacked Abiff. After Abiff refused to divulge the secrets, he is killed. Once King Solomon is told that the Grand Master has disappeared, Solomon orders that all the workers search for the missing Grand Master. Finally, King Solomon is told that three fellow craftsmen, Jubela, Jubelo and Jubelum are missing."[9] The three assassins are eventually captured and killed on orders of King Solomon. "Thus ends the initiation of the Master Mason, most interesting of the three degrees because it contains the unexplained allegory that gave Freemasonry its central identification with the construction of the Temple of Solomon."[10]

As for the three assassins themselves: Perry Raymond Russo told a New Orleans grand jury that CIA agent David Ferrie said (regarding the assassination of JFK) that "there would have to be a minimum of three people involved. Two of the persons would shoot diversionary shots and the third ... shot the good shot." Ferrie said that one of the three would have to be the "scapegoat." He also said that Ferrie discoursed on the availability of exit, saying that the sacrificed man would give the other two, time to escape.[11]

The Warren Commission

"Mason Lyndon Johnson appointed Mason Earl Warren to investigate the death of Catholic Kennedy. Mason and member of the 33rd degree, Gerald R. Ford, was instrumental in suppressing what little evidence of a conspiratorial nature reached the commission. Responsible for supplying information to the commission was Mason and member of the 33rd degree, J. Edgar Hoover. Former CIA director and Mason Allen Dulles was responsible for most of his agency's data supplied to the panel.

"Is it paranoid to be suspicious of the findings of the panel on these grounds? Would it be paranoid to suspect a panel of Nazis appointed to investigate the death of a Jew or to suspect a commission of Klansmen appointed to investigate the death of a negro?

"Dr. Albert Mackey, Mason, member of the 33rd degree, foremost Masonic historian of the nineteenth century, writing in the *Encyclopedia of Freemasonry*, defined Hoodwink as 'A symbol of the secrecy, silence and darkness in which the mysteries of our art should be preserved from the unhallowed gaze of the profane.'

"That is how they see us, as 'profane,' as 'cowans' (outsiders), unclean and too perverted to look upon their hallowed truths. Yes, murder, sexual atrocities, mind control, attacks against the people of the United States, all of these things are so elevated, so lofty and holy as to be beyond the view of mere humans.

Truth or Consequences

"District Attorney for New Orleans, James Garrison, was supported by a 'Truth or Consequences' Club and is alleged to have been an ex-FBI agent and to have been mentally disturbed at one time. Jim Garrison was an outsider in the Secret Society machinations of the FBI and may very well have been pharmacologically or hypnotically induced to set up his ill-fated investigation and the position he acquired in the 'Truth or Consequences Commission.'

"Truth or Consequences, New Mexico, is a town located on the 33rd degree of parallel latitude, and near the same latitude John Fitzgerald Kennedy became an oblation and on the same latitude is the chief Temple on this planet in the minds of sorcerers, namely the Temple of Solomon at Jerusalem, which was once located there and is sworn to be rebuilt on this 33rd degree.

"In a literal, alchemical sense, the Making Manifest of All That is Hidden is the accomplishment of the 3rd Law of the Alchemists and is, as yet, unfulfilled or at least not completed; the other two have been: the creation and destruction of primordial mater (the detonation of the first Atomic Bomb at the Trinity Site, at White Sands, New Mexico, on the 33rd degree of parallel, breaking apart and joining together are the first principles of alchemy. The atomic bomb broke apart the positive and negative (male and female) elements that compose primordial matter, culminating untold thousands of years of alchemical speculation and practice), the Killing of the King (at the Trinity Site, at Dealey Plaza, Dallas, near the 33rd degree of latitude).

"Only the repetition of information presented in conjunction with knowledge of this mechanism of Making Manifest of All That is Hidden provides the sort of boldness and will which can demonstrate that we are aware of all the enemies, all the opponents, all the tricks and gadgetry, and yet we are still not dissuaded, that we work for the truth for the sake of the truth. Let the rest take upon themselves and their children the consequences of their actions."[12]

THE ALCHEMY OF RITUAL MURDER

Michael Hoffman, author of *Secret Societies and Psychological Warfare*, and an expert on the Kennedy assassination explains: "What ought to be unambiguous to any student of mass psychology, is the almost immediate decline of the American people in the wake of this shocking, televised slaughter. There are many indicators of the transformation. Within a year Americans had largely switched from softer-toned, naturally coloured cotton clothing to garish-coloured artificial polyesters. Popular music became louder, faster and more cacophonous. Drugs appeared for the first time outside the Bohemian subculture ghettos, in the mainstream. Extremes of every kind came into fashion. Revolutions in cognition and behaviour were on the horizon, from the Beatles to Charles Manson, from Free Love to LSD.

"The killers were not caught, the Warren Commission was a whitewash. There was a sense that the men who ordered the assassination were grinning somewhere over cocktails and out of this, a nearly-psychedelic wonder seized the American population, an awesome shiver before the realization that whoever could kill a president of the United States in broad daylight and get away with it, could get away with anything.

"A hidden government behind the visible government of these United States became painfully obvious in a kind of subliminal way and lent an undercurrent of the hallucinogenic to our reality. Welcome to Oz thanks to the men behind Os-wald and Ruby.

"There was a transfer of power in the collective group mind the American masses: from the public power of the elected front-man Chief Executive, to an unelected invisible college capable of terminating him with impunity.

"For the first time in their history since the 1826 Masonic assassination of writer William Morgan, Americans were forced to confront the vertiginous reality of a hidden power ruling their world. Sir James Frazer writing in *The Golden Bough: A Study in Magic and Religion*, explains that when the "divine king" is murdered by one who is himself stronger or craftier, those powers of "divinity" which were the king's are "Sympathetically" and "Contagiously" transferred from the vanquished to the victor.

"The entry of this awareness into the subconscious Group 'Dreaming' Mind of the American masses instituted a new simulacrum. The shocking introduction of a diametrically different, new "reality" is a classic scenario of another phase of alchemical programming known to the cryptocracy as 'Clamores.'

"Now the American people were forced to confront a scary alternative reality, the reality of a shadow government, over which they had neither control or knowledge. The shepherding process was thus accelerated with a vengeance. Avant-garde advertising, music, politics and news would hereafter depict (especially in the electronic media) -- sometimes fleetingly, sometimes openly -- a 'shadow side' of reality, an underground amoral 'funhouse' current associated with extreme sex, extreme violence and extreme speed.

"The static images of the suit-and-tie talking heads of establishment religion, government, politics and business were subtly shown to be subordinate to the Shadow State, which the American people were gradually getting a bigger glimpse of out of the corner of their collective eye. The interesting function of this phenomenon is that it simultaneously produces both terror and adulation and undercuts any offensive against it among its percipients, which does not possess the same jump-cut speed and funhouse ambience.

"There is a sense of existing in a palace of marvels manipulated by beautiful but Satanic princes possessed of so much knowledge, power and experience as to be vastly superior to the rest of humanity. They have been everywhere. They have done everything. They run the show that mesmerises us. We are determined to watch it. We are transfixed and desperate to see their newest production, their latest thrilling revelation, even when the thrills are solely based upon the further confirmation of our dehumanization."[13] As author and JFK investigator J.G. Ballard explains: "In this overlit realm ruled by images of the space race and the Vietnam War, the Kennedy assassination and the suicide of Marilyn Monroe, a unique alchemy of the imagination was taking place... The demise of feeling and emotion, the death of affect, presided like a morbid sun over the playground of that ominous decade.

"The role assigned to us is that of zombies called upon by our shadow masters to perform bit parts and act as stock characters in their spectacular show. This mesmerising process produces a demoralised, cynical, double-mind."[14]

Double-mind, as in subliminal messages, as in mysterious synchronicities, as in occult realm which swept through post-war events with a rapier as we looked down the barrel of a gun, exposing the unctuous underbelly of a yet to be identified form. Perception changes ... and we find ourselves back in the cave watching shadows on the wall, believing them to be real.

All the tradition tells us that some "experts" go the wrong way: the "brothers of the left-hand path" or "black magicians." They use their hard-won knowledge and insight to further their own, very personal, and often very sinister, ends. A program of propaganda, mass trauma, celebrity worship and murder ... reconstituting our cultural archetypes into malleable imprints, an esoteric ride that transcends our current slouching shadow play, nervously awaiting the next shock.

RIHANNA

The curtain is drawn and we are back in the real world. The coffin has been removed and the checkered floor scrubbed and polished. Lights, camera, music. So, what does a subliminal message look like? Here is one example. Robyn Rihanna Fenty and better known as simply Rihanna is a Barbadian born R&B recording artist. Her 2007 album, *Good Girl Gone Bad*, which peaked at number two on the Billboard 200 chart, featured five top-ten hits, including three U.S. number one hit singles, amongst them "Umbrella." She has also amassed over a billion views on YouTube, becoming only the fourth artist to do so.[15]

If you pay close attention to her video for "Umbrella," you may notice something very disturbing about it: A digitally mastered image that simply does not fit into the video unless there was an ulterior reason to include it, like a subliminal message, for example. In one of the frames, Rihanna spread her arms, pushing away the rainwater falling on her. As a result, if you freeze-frame it, you can clearly see an image of a goat's head. Or is it Baphomet's head? Are we hallucinating? Is it trick photography? And if it isn't, then, what's it doing there? Perhaps, it was an accident, a quirk of nature. Perhaps, we are making too much of it. Or perhaps, not.

Basic fluid dynamics will tell you that her body and arm movements can not create the liquid formation in the clearly visible form of goat's head. Not only that, but the liquid clearly comes out from the location of Rihanna's womb. Exaggerating? Keep in mind that music videos are created in a fully controlled studio environment. Therefore, if something is in the music video, you can be certain that it was meant to be there.

Let me give you another example. In Eminem's "Mosh" video clip, there is an image of the Eye of Providence, an all-seeing eye that can be traced back to Egyptian mythology and the Eye of Horus. Many interpret it as the all-seeing-eye of secret societies controlling the world from behind the scenes. It's there, painted over an image of American soldiers, with the headline reading "Congress OKs $87 billion for Iraq." Most people understand the Eye of Providence as the All Seeing Eye, the symbol of Illuminati and One World Government. But it is also the symbol of the Supreme Being, the Great Architect of the Universe, because it appears on every U.S. one-dollar bill, above a topless pyramid, a Masonic symbol for the unfinished Temple of Solomon. However, there are other, more sinister interpretations.

THE CULT OF ISIS

The eye of Horus formed part of an evil cult of Isis, "created as a synthetic religion under the Ptolemaic conquerors of Egypt. The cult was created by the priests of the Delphic cult of Apollo, who used preexisting deities of the Egyptian pantheon to construct a new religious cult based chiefly on Babylonian Hellenic models. The creation of the cult of Stoicism by the same priests during the same Ptolemaic period represents the same ideology as the cult of Isis in a semi-religious form.

"The Ptolemaic cults of Isis and the Stoic cults were the ruling pagan cults of Imperial Rome, introduced to Rome by the priests of Apollo. In creating the cult of Isis, the priests of the cult of Apollo at Delphi used the selected names of deities from the Egyptian pantheon to create a cult combining the Babylonian cultism of Apollo with the Orphic and Dyonisian cults of Greece. The orphic (death) cults of Greece were centered around the figure Osiris. The Egyptian goddess Isis was given the legendary and personal properties of the mother of Dionysus, and the oedipal-incestuous mythology around the three combined figures, Osiris, Isis and Horus, was based on the mythology of the Phrygian cult of Dionysus,"[16] or the Roman equivalent, the cult of Bacchus.

There are other examples. In rapper Prodigy's video clip, "The Life," there is a split second image of the 'all-seeing-eye' painted on the wall. What's it doing there and what does it have to do with the song? Nothing. In Lady Gaga's "Alejandro" video clip, the lead male dancer had an imprint of a pyramid and an all-seeing-eye on the back of his jeans jacket, while Lady Gaga was making an all-seeing-eye sign with her right hand while bending over his outstretched leg. Not to be outdone, U2's "Yahweh" is full of Masonic images, from the All-Seeing-Eye on a cross to a pyramid with the Eye of Providence. In Rihanna's "Umbrella," she is walking on a checkered tiled Masonic floor. The Masonic mosaic pavement theme is also present in Jay Sean's "Down" video clip.

We now know that Freemasonry's origins are linked to the once-powerful and wealthy Knights Templar order.[17] The Knights Templar basis for this mosaic pavement is simple. The battle banner of the Knights, the "Beau Séant," was a vertical design consisting of a black block above a white block below. The black block signified the black world of sin the Templar had left behind, and the white block symbolizes the pure life he had adopted as a soldier for Christ.

So, why do secret societies hide things in plain sight? Among many reasons, according to their belief system, to influence world politics at the most urgent time through reverse psychology. Symbols, don't forget, also represent territorial marks of ownership.

By controlling all of the major corporations on Earth and embedding Masonic and occult references in the media, subliminally or otherwise, secret societies are projecting their strength. The power of the image is not only visual. It is a sophisticated tool designed to convey meaning on the one hand and disorient on the other. It is a form of psychological assault, with the number one target to destabilize the victim's rational belief system.

Given that the references make little reasonable sense to most people, we simply tend to assume that the symbols are a coincidence and reject out of hand the existence of the occult. Needless to say Hollywood is the best tool available to project their subliminal message across. In the movie, Angel & Demons, for example, the Illuminati are referred to by name. Hollywood's rendition of the Illuminati theme makes them look very evil and very … outdated, a sect of sorts from the bygone age, clinging on to their ancient rituals and quaint symbols, such as keys and maps. This is a clever tactic as the viewer scoffs at anyone who mentions the Illuminati by name outside of the pre-arranged and controlled Hollywood environment.

Let's take a closer look at the earlier mentioned "Rain Man" and "Umbrella." They may be defined in the following way:

Umbrella
1. Collapsible canopy that protects from the rain or sun
2. Object like an umbrella
3. Something that gives support, protection, or authority

Umbrella in Latin also translate to "shadow" under the influence of "umbra."[18] Umbra can also mean "Phantom" or "Ghost."

Necronomicon

The name "umbra" appears in the *Necronomicon*, which is a "Fictional" occult book of demonology by H.P. Lovecraft. When the word "Rain" is used as an individual's name, it takes on another meaning. The name "rain" means, "abundant blessings from above." By combining the different definitions of the word "rain," we can come up with other possible meanings:

Someone who likes to make it rain (throwing a lot of money away)
Someone who can give you any woman or man you desire
Someone who can make you rich and famous

You might be asking yourself if there is any physical proof of "Rain Man" in Rihanna's video? There is indeed, and your subconscious mind picked it up near the beginning of the video. The image in question is of Rihanna inside a triangle, bent on her knees, head bowed, body leaning forward, arms pulled back over her shoulder blades. However, several things immediately stand out about this frame.

First of all, it appears that Rihanna is bent over, her head bowed low, her face against the floor, her legs are parted and her arms outstretched, angled behind her. The frame is dualistic in nature. It tells two stories at the same time. When you look at her hair, there are only a few strands falling towards the ground. How is this possible if her head is fully bowed? Due to gravity, shouldn't all of her hair be leaning down towards the ground?

Secondly, there are anatomical impossibilities present in the image. The distance between Rihanna's bowed head and shoulders is anatomically disproportionate. Given the relative position of her trapezius muscles and pectoralis major muscles, Rihanna's neck would have to be eleven inches long in order for her head to reach that low. From the position in the frame, her neck would also have to be stretched downward at an anatomically impossible angle.

But this is not all. The fact that we can simultaneously see the trapezius muscles and her pectoralis major muscles indicates that this frame has been doctored, changed or digitally edited. Trapezius muscles and the pectoralis major muscles are posterior and anterior respectively. In other words, they are opposite of each other. One existing at one's back and the other, at one's front. Then, how is it possible that we are able to see both of them simultaneously?

Obviously, the image was altered, changed and adapted to show something different. Look at the image again. Eyes, eyebrows, small forehead, snout, mouth, horns. As a result of this alteration, the image we are actually seeing is of ... a Rain Man, or the Devil.

Why did we not see this in the first place? Because the frame, which creates the subliminal stimuli, is located below an individual's absolute threshold for conscious perception, which makes it only accessible on a subconscious level. Visual stimuli may be quickly flashed before an individual may process them, or flashed and then masked, thereby interrupting the processing. In the movie *Gladiator*, for example, the character Maximus is taken to be killed and the word "Kennedy" flashes on the screen. At the end when Commodus stabs him, the word appears again. But, you missed it, didn't you. Why? Because the frame of film went by so quickly, the naked eye could not register it, but your subconscious mind could have.

In the end, one is left wondering whether it is possible for Hollywood and the music industry to have been used as a tool of secret, special interests, and whether the selection of themes, scripts, actors, studios, producers, directors was, at times, and during certain periods of international or domestic tensions, politically motivated or the result of an intelligence agenda: a far-sighted psychological warfare?

Endnotes

1. "Masonic Symbolism in the Assassination of JFK," James Shelby Downard with Michael A. Hoffman II, http://beforeitsnews.com/alternative/2011/06/masonic-symbolism-in-the-assassinat

2. Ibid

3. James Shelby Downard with Michael A. Hoffman II, *King Kill/ 33º*, 1998, self published

4. Ibid

5. Albert Mackey, H.L. Haywood, *Encyclopedia of Freemasonry*, Kessinger Publishing, 2003, ISBN-13: 978-0766147195

6. Ibid

7. Ibid

8. *Los Angeles Free Press*, Special Report No. I, pg. 16

9. John J. Robinson, *Born in Blood, The Lost Secrets of Freemasonry*, M.Evans Publisher, p.221, 1989

10. John J. Robinson, *Born in Blood, The Lost Secrets of Freemasonry*, M.Evans Publisher, p.223, 1989

11. Quoted by W.H. Bowart in *Operation Mind Control*

12. Sorcery, Sex, Assassination." in Keith, Jim ed. *Secret and Suppressed*. Portland, OR.: Feral House, 1993

13. Michael Hoffman, *Secret Societies and Psychological Warfare*, Independent History and Research, 2001, ISBN-13: 978-0970378415

14. Michael Hoffma Secret Societies and Psychological Warfare," pp. 91-95, Independent History and Research

15. Steven J. Horowitz. "Rihanna Amasses a Billion YouTube Views," www.theboombox.com/2011/03/02, AOL Music. http://www.theboombox.com/2011/03/02/rihanna-amasses-a-billion-youtube-views/

16. Who is behind the Jim Jones Cult, *EIR* Volume 5, Number 46, November 28, 1978, p. 31

17. John J. Robinson, *Born in Blood, The Lost Secrets of Freemasonry*, M.Evans Publisher, 1989

18. *Encarta World English Dictionary*, St. Martin's Press, 1999, ISBN-13: 978-0312222222

CHAPTER 4

The Doors of Perception:
The CIA's Psychedelic Revolution

For the opening of this chapter, I want to refer the reader to an excellent article by Michael Minnicino, "The Frankfurt School and Political Correctness." He writes, "The simmering unrest on campus in 1960 might well too have passed or had a positive outcome, were it not for the traumatic decapitation of the nation through the Kennedy assassination, plus the simultaneous introduction of widespread drug use. Drugs had always been an 'analytical tool' of the nineteenth century Romantics, like the French Symbolists, and were popular among the European and American Bohemian fringe well into the post-World War II period. But, in the second half of the 1950's, the CIA and allied intelligence services began extensive experimentation with the hallucinogen LSD to investigate its potential for social control. It has now been documented that millions of doses of the chemical were produced and disseminated under the aegis of the CIA's Operation MK-ULTRA. LSD became the drug of choice within the agency itself, slowly making its way into academia, and from there into society at large.

"For instance, it was OSS Research and Analysis Branch veteran Gregory Bateson who 'turned on' the Beat poet Allen Ginsberg to an U.S. Navy LSD experiment in Palo Alto, California. Not only Ginsberg, but novelist Ken Kesey and the original members of the Grateful Dead rock group opened the doors of perception courtesy of the Navy."[1] Many young people became "Deadheads" following the Grateful Dead on tour. Veteran investigative journalist Jim Keith writes: "An FBI internal memo from 1968 mentions the employment of the Grateful Dead as an avenue 'to channel youth dissent and rebellion into more benign and non-threatening directions.' [They] performed a vital service in distracting many young persons into drugs and mysticism, rather than politics."[2] The guru of the "psychedelic revolution," Timothy Leary, "first heard about hallucinogens in 1957 from *Life* magazine (whose publisher, Henry Luce, was often given government acid, like many other opinion shapers), and began his career as a CIA contract employee; at a 1977 reunion of acid pioneers, Leary openly admitted, 'everything I am, I owe to the foresight of the CIA.'"[3]

In his book, titled *Understanding Understanding*, Dr. Humphry Osmond wrote "the introduction of the principal psychedelic drugs in the 1960s was a result of the CIA's investigation into its possible military use."[4] The experiments, under different code names, were conducted at more than eighty university campuses and involuntarily helped to popularize LSD. Thousands of university students were used as unwitting Guinea pigs. Soon they [the students] started synthesizing their own acid.

Beginning in 1962, the RAND Corporation National Defense Research Institute[5] of Santa Monica, California, also began a secret four-year study in experimental psychosis as a tool for psychiatric evaluation using LSD, peyote, and marijuana. RAND,[6] an outgrowth of the wartime Strategic Bombing Survey, a 'cost analysis' study of the psychological effects of the bombings of German population centers, is a federally-funded 'think-tank' sponsored by the Office of the Secretary of Defense and the Council on Foreign Relations. The study was based on the work of psychiatrist W.H. McGlothlin, who conducted preparatory research on the "The Long-Lasting Effects of LSD on Certain Attitudes in Normals: An Experimental Proposal."[7]

Michael Minnicino explains this phenomenon in *Fidelio Magazine*'s, winter 1992 issue: "Hallucinogens have the singular effect of making the victim asocial, totally self-centered, and concerned with objects. Even the most banal objects take on the 'aura' which Benjamin had talked about, and become timeless and delusionarily profound. In other words, hallucinogens instantaneously achieve a state of mind identical to that prescribed by the Frankfurt School theories. And, the popularization of these chemicals created a vast psychological lability for bringing those theories into practice. Thus, the situation at the beginning of the 1960's represented a brilliant re-entry point for the Frankfurt School, and it was fully exploited. One of the crowning ironies of the 'Now Generation' of 1964 on, is that, for all its protestations of utter modernity, none of its ideas or artefacts was less than thirty years old.

"The political theory came completely from the Frankfurt School; Lucien Goldmann, a French radical who was a visiting professor at Columbia University in 1968, was absolutely correct when he said of Herbert Marcuse in 1969 that 'the student movements ... found in his works and ultimately *in his works alone* the theoretical formulation of their problems and aspirations [emphasis in original].' The long hair and sandals, the free love communes, the macrobiotic food, the liberated lifestyles, had been designed at the turn of the century, and thoroughly field-tested by various, Frankfurt School-connected New Age social experiments like the Ascona commune before 1920. Even Tom Hayden's defiant 'Never trust anyone over thirty,' was merely a less-urbane version of Rupert Brooke's 1905, 'Nobody over thirty is worth talking to.' The social planners who shaped the 1960's simply relied on already-available materials, taking the entire charade one step further, beyond sanity and into the world of super-drugs to be used for behaviour modification and mind control tasking, a key that would eventually allow unfettered access to the human mind."[8]

To understand the roots of this charade, Jeffrey Steinberg of the *Executive Intelligence Review* suggest that it is necessary to go back to the immediate postwar period, "when there was a concerted effort launched, by the Frankfurt School and the London Tavistock Institute, to use the Marxist/Freudian perversion of psychology and other social sciences, as instruments for mass social control and brainwashing. The two pillars of the assault on the American intellectual tradition were cybernetics and the drug counterculture.

"Drug counterculture was precisely the weapon that the Frankfurt School and their fellow-travellers employed, over the next 50 years, to create a cultural paradigm shift away from the so-called 'authoritarian' matrix of man in the living image of God, and the superiority of the republican form of nation-state over all other forms of political organization. They transformed American culture toward an erotic, perverse matrix, associated with the present 'politically correct' tyranny of tolerance for dehumanizing drug abuse, sexual perversion, and the glorification of violence. For the Marxist/ Freudian revolutionaries of the Frankfurt School, the ultimate antidote to the hated Western Judeo-Christian civilization was to tear that civilization down, from the inside, by turning out generations of necrophiliacs.

"If this statement seems harsh, consider the following. In his 1948 work on *The Philosophy of Modern Music*, Frankfurt School leader Theodor Adorno argued that the purpose of modern music is to literally drive the listener insane. "He justified this by asserting that modern society was a hotbed of evil, authoritarianism, and potential fascism, and that, only by first destroying civilization, through the spread of all forms of cultural pessimism and perversity, could liberation occur. On the role of modern music, he wrote, 'It is not that schizophrenia is directly expressed therein; but the music imprints upon itself an attitude similar to that of the mentally ill. The individual brings about his own disintegration.... He imagines the fulfilment of the promise through magic, but nonetheless within the realm of immediate actuality.... Its concern is to dominate schizophrenic traits through the aesthetic consciousness. In so doing, it would hope to vindicate insanity as true health.' Necrophilia, he added, is the ultimate expression of 'true health' in this sick society.

"Erich Fromm, another leading Frankfurt School figure, devoted much of his seminal 1972 work, *The Anatomy of Human Destructiveness*, to the analysis of necrophilia, which he pronounced to be the dominant trend in modern society. Fromm defined necrophilia as all forms of obsession with death and destruction, particularly those with intense sexual overtones. Ironically, his ostensible 'cure' for this mass social perversion was the drug, rock, sex counterculture of the late 1960s."[9]

LIBERATION THROUGH DRUG ABUSE

Earlier, in 1952, a proposal was issued to the Director of Central Intelligence outlining funding mechanisms for highly sensitive CIA research and development projects that would study the use of biological and chemical materials such as LSD, in altering human behaviour. On April 13, 1953, MK-ULTRA, was established for the express purpose of researching and developing chemical, biological, and radiological materials to be used in clandestine operations and capable of controlling or modifying human behaviour.

Again, quoting Jeffrey Steinberg, "As these monstrous notions of mass social engineering were being presented as the 'humanistic' alternative to world war in the age of the atomic and hydrogen bomb, crucial projects were being launched, that would shape the implementation of this Brave New World, and bring us, today, kicking and screaming to the world of the Aquarian Conspiracy. It is noteworthy that one of the four directors of the Authoritarian Personality project, R. Nevitt Sanford, played a pivotal role in the 1950s' and '60s' experimentation and eventual mass usage of psychedelic drugs."[10]

The Authoritarian Personality project of the late 1940s, was responsible for engineered the Baby Boomer drug/rock/sex counterculture of the 1960s. "In 1965, Sanford wrote the forward to *Utopiates: The Use and Users of LSD 25*, which was published by Tavistock Publications, the publishing arm of Great Britain's pre-eminent psychological warfare agency, the Tavistock Institute. Tavistock directed the Psychiatric Division of the British Army during World War II, and dispatched many of its top brainwashers to the United States in the immediate postwar period, to work on the secret mind-control projects of the CIA and the Pentagon, including the MK-ULTRA project, devoted to the study of LSD and other psychedelics. In his foreword to *Utopiates*, Sanford, who headed up the Stanford University Institute for the Study of Human Problems, a major outpost for MK-ULTRA secret LSD experimentation."[11]

'The nation,' Sanford wrote, "'seems to be fascinated by our 40,000 or so drug addicts who are seen as alarmingly wayward people who must be curbed at all costs by expensive police activity. Only an uneasy Puritanism could support the practice of focusing on the drug addicts (rather than our 5 million alcoholics) and treating them as a police problem instead of a medical one, while suppressing harmless drugs such as marijuana and peyote along with the dangerous ones.'"[12]

Lysergic acid diethylamide, or LSD, was developed in 1943 by Albert Hoffman, a chemist at Sandoz A.B. – a Swiss pharmaceutical house owned by S.G. Warburg. LSD's principal component is ergotamine tartrate, which Hoffman turned into synthetic ergotamine, a powerful mind altering, and highly addictive drug. Looking deeper into the LSD history we discover that "While precise documentation is unavailable as to the auspices under which the LSD research was commissioned, it can be safely assumed that British intelligence and its subsidiary U.S. Office of

Strategic Services were directly involved. Allen Dulles, the director of the CIA when that agency began MK-ULTRA, was the OSS station chief in Berne, Switzerland throughout the early Sandoz research. One of his OSS assistants was James Warburg, of the same Warburg family, who was instrumental in the 1963 founding of the Institute for Policy Studies, and worked with both Huxley and Robert Hutchins."[13]

All in all, there were 149 MK-ULTRA subprojects, many of them involved with research into behaviour modification, hypnosis, drug effects, psychotherapy, truth serums, pathogens and toxins in human tissues.

For intelligence purposes, what was required of MK-ULTRA was the ability to manipulate memory. Thus, what the Freudians call the "superego," Peter Levenda explains in Sinister Forces, "had to be bypassed, allowing the controller direct access to the contents of an enemy agent's mind. That was step one. Step two would involve erasing specific pieces of information from the subject's memory and replacing those pieces with new bits of memory, thus permitting the Agency to send that agent back into the field without any knowledge that he or she had been interrogated and had given up sensitive information. Step three was a potential bonus: Could that enemy agent then 'be programmed' to commit acts on behalf of the Agency, without knowing who gave the command or why? This was the essence of the Manchurian Candidate. It is also the essence of what we know today as hypnotherapy and 'depth' psychoanalysis, for the psychiatrist is looking for access to the patient's unconscious layers, to extract important information, such as childhood trauma, and to neutralise the effects of that trauma, in some cases replacing certain behaviour patterns with new, approved patterns."[14]

The proximate origins for these actions "go back to an overtly satanic turn in the Nineteenth-Century Romantic movement. Notably, the exemplary connection of Britain's H.G. Wells, Bertrand Russell, and Theosophist leader Aleister Crowley, to the role of Aldous Huxley in the promotion of what were originally known as psychotomimetic ("psychosis-imitating"), or psychedelic drugs. This connection exposes the satanic quality of the work of Aldous Huxley et al. in the preparation of what emerged as the "rock-drug-sex counterculture" of the 1960s."[15]

Crowley and Capri resident Aksel Muenthe were the leading figures of the Satan/anti-Christ cult of Theosophy during the beginning of 20th century. "The publication of the pro-satanic periodical, *Lucifer*, in Vienna, with participation of Crowley, is exemplary. The role of Maxim Gorky in the Grotto of Capri, was part of the pre-World War I activity of these pro-satanic circles. The composer Richard Wagner and the existentialist Friedrich Nietzsche, were key figures of this anti-Christ movement. For Muenthe et al., the original anti-Christ was the Emperor Tiberius whose son-in-law, Pontius Pilate, executed Jesus Christ."[16] This was the circle into which Aldous and Julian Huxley were introduced, by H.G. Wells and Bertrand Russell, during the late 1920s and early 1930s. "Notably, the Josiah Macy, Jr. Foundation of the 1940s and 1950s served as one of the principal covens where the architects of the 1960s rock-drug-sex youth-counterculture often met."[17]

MASS PSYCHOLOGY MODEL

The counterculture model that the British had imposed on the United States was based on the pagan ceremonies of ancient Egyptian and Roman Empires. And these had a history of their own. The continuity of the cult of Apollo is important to this point. There are "black nobility" families of Rome whose families and family political traditions trace back to the Roman republic. "The republic and empire under which their ancestors lived was in turn controlled by Rome branch of the cult of Apollo. That cult, was during that time, variously, the chief usurious debt-farming institution of the Mediterranean region,"[18] a political intelligence service, and both a cult and a creator of cults.

From the death of Alexander the Great until the cult of Apollo dissolved itself into the cult of Stoicism it had created during the second century B.C., the base of the cult was Ptolemaic Egypt, from which the cult controlled Rome. In Egypt, the cult of Apollo syncretized the cult of Isis and Osiris as the direct imitation of the Phrygian cult of Dionysus and the Roman imitation, the cult of Bacchus. It was there that the cult of Stoic irrationalism was created by the cult of Apollo. It was the cult of Apollo that created the Roman Empire, which created the Roman law on the basis of the antihumanistic Aristotelean Nicomachean Ethics. That is the tradition which the old 'black' Roman families transmit. These roman families, over time, became known as the Venetian Black Nobility and today their members hold key positions within the inner circle of organizations such as the Bilderberg group.

That tradition persisted under various institutional covers, always preserving the essential world-outlook and doctrine intact. The British monarchy, the parasitical aristocratic landlord class of Britain, and the British-dominated, feudalist factions of the Maltese Order, are the modern, concentrated expression of the unbroken tradition and policies of the ancient Apollo cult.

The Aristotelian knows that "generalised scientific and technological progress, given the conditions of education and liberty of innovation which progress demands, produces in the citizen a dedication to the creative potential of the human mind which is antithetical to the oligarchical system.

What the Aristotelians have hated and feared down through the millennia is their knowledge that persistent, generalized scientific and technological progress, as the dirigiste policy of society, means a republican hegemony which ends forever the possibility of establishing oligarchical world-rule."[19]

They have resorted to the same methods used by the ancient priests of Apollo – the promotion of Dionysian cults of drug-cultures, orgiastic-erotic countercultures, deranged mobs of "machine-breakers" and terrorist maniacs – to turn such a combined force of demented rabble against those forces in society that are dedicated to scientific and technological progress.

The following description of cult ceremonies dating back to the Egyptian Isis priesthood of the third millennium B.C. could just as well be a journalistic account of a "hippie be-in" circa A.D. 1969: "The acts or gestures that accompany the incantations constitute the rite (of Isis). In these dances, the beating of drums and the rhythm of music and repetitive movements were helped by hallucinatory substances like hashish or mescal; these were consumed as adjuvants to create the trance and the hallucinations that were taken to be the visitation of the god. The drugs were sacred, and their knowledge was limited to the initiated.... Possibly because they have the illusion of satisfied desires, and allowed the innermost feelings to escape, these rites acquired during their execution a frenzied character that is conspicuous in certain spells: "Retreat! Re is piercing thy head, slashing thy face, dividing thy head, crushing it in his hands; thy bones are shattered, thy limbs are cut to pieces!"[20]

Firstly, Isis is a drug cult. It is a cult of high priests and secret rituals. These secret rituals have been kept inside the British Royal Family and their ruling class friends, for centuries. The Cult was practiced in Egypt during the Third Dynasty of the Old Kingdom circa 2780 B.C. Isis worship is essentially pagan, primitive mother worship. The Isis priesthood is a closed circle and its nobility practiced social control, total subjugation of man's free will, exploitation and subjugation of people. The Isis Cult was "popularised" minus its secrets, by Isis high priest Bulwer-Lytton's work, *The Last Days of Pompeii*. Bulwer-Lutton's son, Edward, was Viceroy and Governor General of India from 1876 through 1880, when exports of Bengal opium to China were greatly stepped up. Bulwer-Lutton was the mentor of Lord Palmerston who led the British Parliament during the opium wars that forced an unwilling China, to not only continue, but also expand the sale of opium in that country. Russian occultist Madame Blavatsky further popularized the Cult of Isis through her book, *Isis Unveiled*.

Lord Bertrand Russell, who joined with the Frankfurt School in this effort at mass social engineering for the purposes of destroying man's creative powers as well as wreck European and American cultures, spilled the beans in his 1951 book, *The Impact of Science on Society*. He wrote:

"Physiology and psychology afford fields for scientific technique which still await development. Two great men, Pavlov and Freud, have laid the foundation. I do not accept the view that they are in any essential conflict, but what structure will be built on their foundations is still in doubt. I think the subject which will be of most importance politically is mass psychology.... Its importance has been enormously increased by the growth of modern methods of propaganda. Of these the most influential is what is called 'education.' Religion plays a part, though a diminishing one; the press, the cinema, and the radio play an increasing part.... It may be hoped that in time anybody will be able to persuade anybody of anything if he can catch the patient young and is provided by the State with money and equipment."

Russell continued, "The subject will make great strides when it is taken up by scientists under a scientific dictatorship . . . The social psychologists of the future will have a number of classes of school children on whom they will try different methods of producing an unshakable conviction that snow is black. Various results will soon be arrived at. First, that the influence of home is obstructive. Second, that not much can be done unless indoctrination begins before the age of ten. Third, that verses set to music and repeatedly intoned are very effective. Fourth, that the opinion that snow is white must be held to show a morbid taste for eccentricity. But I anticipate. It is for future scientists to make these maxims precise and discover exactly how much it costs per head to make children believe that snow is black, and how much less it would cost to make them believe it is dark gray."

Russell concluded with a warning: "Although this science will be diligently studied, it will be rigidly confined to the governing class. The populace will not be allowed to know how its convictions were generated. When the technique has been perfected, every government that has been in charge of education for a generation will be able to control its subjects securely without the need of armies or policemen."[21]

The Russells, an English noble family, came into their own during the reign of Henry VIII. Earl Bertrand Russell was the grandson of Lord John Russell (1798-1878), twice prime minister during the reign of Queen Victoria.

ENTER ALDOUS HUXLEY

It was Lord Russell who proposed the development of mass drugging of populations as method of social control, through decriminalization of marijuana and other dangerous psychotropic drugs and the modification of the use of language to provide methods of mass social manipulation (linguistics). The drug side of the Russell project was picked up chiefly by Aldous Huxley.

Aldous Huxley was a grandson of the famous biologist, Thomas Henry Huxley, whose outspoken support of Charles Darwin's theory of evolution earned him the nickname, 'Darwin's Bulldog.' It was Thomas Huxley, "a leading member of the Metaphysical Society, which was founded in 1869 in an attempt to forge a more effective intellectual elite out of the membership of the Oxford Essayists and Cambridge Apostles"[22] who coined the term 'agnosticism.' Huxley and his leading circles denied the ability of man to actually know anything, the so-called "no-soul" doctrine that was at the heart of Wells' book, *Open Conspiracy*.

Unlike his grandfather, Aldous Huxley was a novelist, who liked to experiment with drugs to discover his inner creativity and spiritual connection. Drugs, more than evangelists, he believed, had the psychedelic power to tap into a spiritual well of visions. Even a fleeting experience of self-transcendence could shatter society's traditional way of thinking about religion in each person's pursuit of a deeper and more meaningful spiritual life. Huxley predicted that religion would be transformed from "an activity mainly concerned about symbols" to "an activity mainly concerned with experience and intuition" driven by mystical inspiration.

It all went back to Thomas Huxley's basic teaching: "the unifying theme was the denial of the nobility of man, as expressed in the provable power of the human mind to create and discover new ideas."[23]

Aldous Huxley was a lifelong collaborator of Arnold Joseph Toynbee, an economic historian whose twelve-volume analysis of the rise and fall of civilizations examined history from a global perspective. Huxley met Toynbee at Oxford, where he was tutored by H.G. Wells, who was T.H. Huxley's protégé. While Wells went on to become the head of British Foreign Intelligence during World War I, Toynbee sat on the Royal Institute of International Affairs (RIIA) Council in Chatham House [interlocked with the CFR in the United States] for nearly fifty years. During World War I, he also headed the Research Division of the Intelligence Department for the British Foreign Office and was a delegate to the Paris Peace Conference in 1919.[24] From 1925 to 1955, he served as the Director of Studies at the RIIA, and then during World War II, he became Prime Minister Winston Churchill's briefing officer.

In an excellent investigative report on the Aquarian Conspiracy, *Executive Intelligence Review* staff explain that Toynbee's "theory" of history, "expounded in his twenty-volume History of Western civilization, was that its determining culture has always been the rise and decline of grand imperial dynasties. At the very point that these dynasties – the 'thousand year Reich' of the Egyptian pharaohs, the Roman Empire, and the British Empire – succeed in imposing their rule over the entire face of the earth, they tend to decline. Toynbee argued that this decline could be abated if the ruling oligarchy (like that of the British Roundtable) would devote itself to the recruitment and training of an ever-expanding priesthood dedicated to the principles of imperial rule."[25]

Whereas Toynbee appealed to the high-brow intellects of the British aristocracy, Wells' science fiction stories had squarely positioned him as a 'pop star' of the time. According to Wells, world revolution was only attainable through an *Open Conspiracy: Blue Print for a World Revolution*, using counterculture as a battering ram on an unsuspecting public. Wells wrote, "[it] will appear first, I believe, as a conscious organization of intelligent and quite possibly in some cases, wealthy men, as a movement having distinct social and political aims, confessedly ignoring most of the existing apparatus of political control, or using it only as an incidental implement in the stages, a mere movement of a number of people in a certain direction who will presently discover with a sort of surprise the common object toward which they are all moving ... In all sorts of ways they will be influencing and controlling the apparatus of the ostensible government."[26]

From his experience as one of the initiates in the "Children of the Sun," a Dionysian cult comprised of the children of Britain's Round Table elite,[27] Aldous Huxley gained material to create his most famous novel, *Brave New World* [28] published in 1931. *Brave New World* is a virtual blueprint of a true-life, future, one-world socialist government, or as his Fabian mentor, H.G. Wells, paraphrased the title for one of his own popular novels, *The New World Order* [1940], *The Open Conspiracy: Blue Prints for a World Revolution* [1928], which openly examines how this New World Order could be attained, ostensibly for world peace and human evolution.

Remember, the goal of Wells, Russell, Huxley and company is the destruction of the sovereign power of the nation-state, and with it the elimination of a philosophical, cultural, and religious tradition dating more than 2,500 years.

So, what is it that makes an 'open conspiracy' open, as in placed there in plain sight for everyone to see? Larry Hecht explains: "what makes the 'open conspiracy' open, is not the laying out of some secret masterplan, not the revealing of the membership roster of some inner sanctum of the rich and powerful, which the typical deluded populist supposes to be the secret to power in the world.

It is, rather, the understanding that ideas, philosophy and culture, control history. What constitutes a conspiracy, for good or evil, is a set of ideas which embody a concept of what it is to be human, and a conception of man's role in universal history."[29]

It was H.G. Wells who introduced Huxley to Satanist Aleister Crowley. Crowley was a poster child of the occult that developed in Britain from the 1860s under the influence of Edward Bulwer-Lytton – the colonial minister under Lord Palmerston during the Second Opium War. In 1886, Crowley, William Butler Yeats, poet and Nobel Prize winner and several other Bulwer-Lytton proteges formed a satanic "Isis-Urania Temple of Hermetic Students of the Golden Dawn. This Isis Cult was organized around the 1877 manuscript *Isis Unveiled* by Madame Helena Blavatsky, in which the Russian occultist called for the British aristocracy to organize itself into an Isis priesthood."[30]

Criton Zoakos explains that "among the other initiates *in the Children of the Sun* were T.S. Eliot, W.H. Auden, Sir Oswald Mosley, and D.H. Lawrence, Huxley's homosexual lover. It was Huxley, furthermore, who would launch the legal battle in the 1950s to have Lawrence's pornographic novel *Lady Chatterley's Lover* allowed into the United States on the ground that it was a misunderstood 'work of art.'"[31]

In *Brave New World*, Huxley centers on the scientific methodology for keeping all populations outside the elite minority in a permanently autistic-like condition actually in love with their servitude. Speaking at the California Medical School in San Francisco, Huxley announced: "There will be in the next generation or so a pharmacological method of making people love their servitude and producing dictatorship without tears, so to speak. Producing a kind of painless concentration camp for entire societies so that people will in fact have their liberties taken away from them but will rather enjoy it, because they will be distracted from any desire to rebel by propaganda, or brainwashing, or brainwashing enhanced by pharmacological methods. And this seems to be the final revolution." To Wells' way of thinking, this wasn't a conspiracy, but rather a *one-world brain* functioning as *a police of the mind.*

In a speech on the U.S. State Department's Voice of America, in 1961, Huxley spoke of a world of pharmacologically manipulated slaves, living in a "concentration camp of the mind," enhanced by propaganda and psychotropic drugs, learning to 'love their servitude,' and abandoning all will to resist. "This," Huxley concluded, "is the final revolution." Huxley's cohort in the 1950s experimentation with psychotropic drugs, Dr. Timothy Leary, of Harvard University's Psychology Department, "provided another glimpse into the perverted minds of the Russell/Huxley/Frankfurt School crowd, in *Flashback*, his autobiographical account of the Harvard University Psychedelic Drug Project. Leary quoted Huxley: 'These brain drugs, mass produced in the laboratories, will bring about vast changes in society. This will happen with or without you or me. All we can do is spread the word. The obstacle to this evolution, Timothy, is the Bible.' Leary then added: 'We had run up against the Judeo-Christian commitment to one God, one religion, one reality, that has cursed Europe for centuries and America since our founding days. Drugs that open the mind to multiple realities inevitably lead to a polytheistic view of the universe. We sensed that the time for a new humanist religion based on intelligence, good-natured pluralism and scientific paganism had arrived.'"[32]

HUXLEY AT WORK

In 1954 Huxley published a study of consciousness expansion through mescaline, *The Doors of Perception*, the first manifesto of the psychedelic drug cult. The Doors[33] rock band borrowed their name from this psychedelic manifesto. Huxley borrowed the title from William Blake's poem: "If the doors of perception were cleansed / All things would appear infinite."

In 1958, Aldous Huxley assembled a number of essays that he had written for *Newsday* and published them collectively as *Brave New World Revisited*, in which he described a society where "the first aim of the rulers is, at all costs, to keep their subjects from making trouble." He described a likely future: "The completely organized society ... the abolition of free will by methodical conditioning, the servitude made acceptable by regular doses of chemically induced happiness..."

He also predicted democracies would change their nature: "quaint old forms – elections, parliaments, Supreme Courts – will remain but the underlying substance will be *non-violent totalitarianism*. Democracy and freedom will be the theme of every broadcast and editorial – but democracy and freedom, in a strictly Pickwickian [i.e. not literal] sense. Meanwhile, the ruling oligarchy and its highly trained elite of soldiers, policemen, thought-manufacturers and mind-manipulators will quietly run the show as they see fit."[34]

How different is it from what we have today?

THE ROOTS OF THE FLOWER PEOPLE

Back in California, Gregory Bateson had maintained the Huxley operation out of the Palo Alto VA hospital. Bateson, one of the most colourful characters of the epoch, was former husband of anthropologist Margaret Mead. Bateson, himself a renowned anthropologist with the OSS, "became the director of a hallucinogenic drug experimental clinic at the Palo Alto Veterans Administration Hospital. Under Bateson's auspices, the initiating "cadre" of the LSD cult – the hippies – were programmed.

"Through LSD experimentation on patients already hospitalized for psychological problems, Bateson established a core"[35] of 'initiates' into the 'psychedelic' Isis Cult. Here is a bit of what Michael Minnicino in the *The Campaigner's* April 1974 issue had to say: "Foremost among his Palo Alto recruits was Ken Kesey. In 1959, Bateson administered the first dose of LSD to Kesey. By 1962, Kesey had completed a novel, *One Flew Over the Cuckoo's Nest*, which popularized the notion that society is a prison and the only truly 'free' people are the insane."[36] It is worth noting that during the 1960s, the Tavistock Clinic fostered the notion that no criteria for sanity exist and that psychedelic "mind-expanding" drugs are valuable tools of psychoanalysis. Kesey subsequently organized[37] a circle of LSD initiates' called 'The Merry Pranksters.'"[38]

This goofy band toured the United States spreading LSD and "establishing the pretext for a high volume of publicity on behalf of the still minuscule"[39] counterculture.

Minnicino's report continued: "By 1967, the Kesey cult had handed out such quantities of LSD that a sizable drug population 'flower children' had emerged, centerd in the Haight-Ashbury district of San Francisco. Here Huxley collaborator Bateson set up a 'free clinic,' staffed by Dr. David Smith, later a 'medical adviser' for the National Organization for the Reform of Marijuana Laws (NORML) and Dr. Peter Bourne, formerly President Carter's special assistant on drug abuse."

PAGAN CONNECTION

Yet, it was not until the Vietnam War and the "anti-war" movement that really created the climate of moral despair and rot that spurred America's youth to rely on drugs. In other words, the war was used as a backdrop against which to create the anti-war movement. Their disillusionment with the war made the protestors easy prey for their masters to condition them under the guidance of the Tavistock Institute and the CIA through the influence of the LSD drug culture.

I am not suggesting that most anti-war protesters were paid agents for the elite circles. On the contrary, as Criton Zoakos suggests, "the overwhelming majority of anti-war protesters went into the Students for a Democratic Society (SDS) organization on the basis of outrage at the developments in Vietnam. But once caught in the environment defined by the Tavistock Institute's psychological warfare experts, and inundated with the message that hedonistic pleasure-seeking was a legitimate alternative to 'immoral war,' the protesters' sense of values and their creative potential went up in a cloud of hashish smoke.

"The outcome of this debacle was a major strategic withdrawal from Asia by the United States, spelled out in Henry Kissinger's 'Guam Doctrine,' adoption of the spectacular failure known as the 'China Card' strategy for containing Soviet influence, and demoralization of the American people over the war to the point that the sense of national pride and confidence in the future progress of the republic was badly damaged. "[40]

That moral despair is what the Frankfurt School had hoped for back when they were looking for new cultural forms to increase the alienation of the population. They found it in the newly created Rockefeller-Tavistock sponsored and financed organization, IPS, Institute for Policy Studies.

INSTITUTE FOR POLICY STUDIES

Established in 1963, IPS was run by President Kennedy's National Security Advisor and former President of Ford Foundation, McGeorge Bundy, one of the architects of U.S. policy toward Vietnam, particularly the notorious Strategic Hamlet programs and the murderous Operation Phoenix program in the Mekong Delta region.

IPS's main assignment was to control and co-ordinate a wide array of pseudo-organizations, from local community control and black nationalist groups to anti-technology organizations, anti-war movement and terrorist operations such as the Weathermen, especially under the auspices of the anti-war movement, to zero growth and environmentalist freak shows – "operations of the sort which were pioneered by British intelligence agencies, which place major emphasis upon the use of private institutions whose overall activity is inextricably linked to official government institutions."[41] In short, the entirety of what became known in the 1960s as the 'New Left Movement.' In their well-laid-out deconstruction of Georgii Arbatov's book, *The War of Ideas in Contemporary International Relations*, the investigative team of *EIR* writes, "The FBI and sections of the CIA have provided the controllers and deployment managers of the 'New Left.' IPS has provided the 'cover stories' and the 'ideological management.'"[42]

Chief theoretician for the IPS was none other than Noam Chomsky, one of the founders of the New Left. For Chomsky, the primary purpose for creating the New Left of the early 1960s was to pre-empt the radicalization of college youth strata to two overlapping ends. Immediately, the purpose was to prevent the established socialist parties from effectively capturing the social ferment, which had erupted beginning in 1958, "which had come to focus around the Cuban Revolution and Civil Rights movement up through mid-1961. At the same time, with such enterprises as the Peace Corps, Office of Economic Opportunity and foundation-sponsored *community action projects for radicals* in development at that time, to drain off student radicalism into staffing of an expanding counterinsurgency apparatus around the Reesian fascist conception of *local community control*."[43]

In an exclusive report on the Institute for Policy Studies, the investigative team from Executive Intelliegence Review points out that "It is important to note that these operations are the outcome of political intelligence operations set up prior to World War I under the auspices of the National Civic Federation and the Russell Sage Foundation and which were officially institutionalized following World War II under the aegis of the CIA."[44]

The Institute for Policy Studies lists, among its lecturers and fellows, members of the Weathermen group: a U.S. radical-left organization consisting of splintered-off members and leaders of the Students for a Democratic Society. In anti-war activities, IPS members carried out a series of militant actions, including bombings and violent protests to achieve the overthrow of the U.S. Government. This group was also part of the counter-counterculture controlled from within by the U.S. Government.

The Cambridge Institute in Cambridge, Massachusetts, the Transnational Institute in Amsterdam, Holland, and related organizations "such as the group of so-called ex-CIA and military intelligence personnel around CounterSpy promote such policies to seed the political climate with the idea that many of these actually covert operations are really expressions of popular political will. With the help of the mass media and selected newspapers, many of the operations were made to appear as legitimate expressions of a political point of view." [45]

FUNDING OF IPS AND RELATED INSTITUTIONS

The foundations that provide the money to maintain the illegal activities of the Institute can be divided into two basic categories. First, top level foundations all linked to the Rockefeller family such as the Rockefeller Family Fund, Rockefeller Brothers Fund, Ford Foundation, serve as organizations that disburse funds for the specific operations. What these 'specific operations' are, we really do not know, as the official reports never identify their purpose; instead a 'cover' organization or activity is used to facilitate the disbursement of funds. On the second level, foundations, such as the Kaplan Fund and Stern Fund, serve as conduits for political and covert functions.

THE CASE OF MARCUS RASKIN

The key individual involved with IPS was Marcus Raskin, intimately linked to the top levels of Rockefeller's political intelligence machine. Raskin's earlier training came at the University of Chicago Law School, a political factory for Rockefeller agents and operatives.

By the mid-1960s, Rockefeller and the founders of the Institute for Policy Studies, Hans Morgenthau, a Rockefeller man and one of the founding fathers of the "realist school" of the 20th century and McGeorge Bundy, began to establish a cover for running anarchist-leftist operations via the co-opting and influencing of a vast international network of operatives for 'black operations.' This was done through the National Security Council's secret unit known as "Special Staff" that plan and coordinate the psycho-political operations used to manipulate the American public, and through a vast intergovernmental undercover infrastructure that include the legislative, executive and judicial branches of government, such as the Secretary of State, the Secretary of Defense, the Secretary of the Treasury and the Director of the CIA; those who control television, radio, and newspaper corporations; who head the largest law firms; who run the largest and most prestigious universities and think-tanks; who direct the largest private foundations and who direct the largest public corporations. Within the Special Staff, Raskin became one of the overseers of special covert operations and destabilization scenario planning – a unique experience in creating and developing terror.

NSC Special Staff, on the one hand, was a part of strategic psycho-political operations that focus propaganda at small groups of people, such as academics or experts, capable of influencing public opinion and on the other, a part of tactical psycho-political operations that focus propaganda at the masses through mass-communication media.

A vital part of the operation was the creation of the umbrella of the student anti-war movement, Students for a Democratic Society in the U.S. With the help of the leading U.S. figures in the social democratic "Second International," such as Michael Harrington and UAW leaders Victor and Walter Reuther. The SDS, created by the League for Industrial Democracy, became the flagship "synthetic new left" organization for spinning off phony 'leftist' operations internationally. In turn, the League of Industrial Democracy was financed and run by the Institute for Policy Studies. "Through a process of splitting and then regrouping such organizations, a selection procedure was established for creating terrorist organizations. Raskin supervised this operation starting in 1963 and including the period when the Weathermen and other terrorist organizations were set up."[46]

As a co-director of IPS, Raskin initiated a series of projects designed for the profiling and eventual deployment of the various re-groupments of SDS and left groups internationally. "He supervised the Radical Education Project, created in part by the University of Michigan's Institute for Social Research, the U.S.'s leading 'conflict resolution' think-tank."[47] Raskin had also been thoroughly trained in psychological profiling through the National Training Laboratories, a project set up for the Rockefellers by the German fascist psychologist Kurt Lewin. NTL specializes in brainwashing through various methods, "including transactional analysis and therapy and group therapy, as well as harder brainwashing methods, all employed to create the social basis for running terrorist operations."[48]

For example, the "second generation of the Baader-Meinhof gang came from the graduates of the Heidelberg Patients' Collective, an experimental program using mental patients, in which the patients were indoctrinated according to R.D. Laing-Tavistock schizophrenia-reenforcing program and also, as part of the program, assigned to develop the capability to produce bombs."[49]

It was the same technique used by Francesca Alberoni, the University of Trento's School of Sociology head, to create Italy's "leftist" terrorist organization the Red Brigades. The Trento program was a social-work brainwashing program. The same modus operandi was used to create the Weathermen Underground as a terrorist group.

"Not only was there irrationalist anarchist indoctrination, largely under the guidance of IPS, but 'sensitivity sessions', degraded sexual practices, and emphasis on the use of LSD and other drugs were dominant features of the conditioning process. The rock-drug counterculture, including sexually and otherwise degraded communal living conditions, is the general rule among the strata from which terrorists are recruited."[50]

This is the true base of the "radical youth idealism" of the 1960s, known as the counterculture movement, that spawned the Weathermen and every extreme left-wing organization in the United States. Under the auspices of Tavistock Institute and the assorted support structure, such as IPS, the indoctrination terrorist program functioned in the following manner:

The intake process began with a leftist-environmentalist profiling and recruitment under the auspices of Amnesty International. It would last several weeks, with drugs and wife swapping, apart from "leaderless group" approaches to deeper political indoctrination. The second principal stage involved a months-long period, augmented with drug use and sexual degradation.

The third stage was Longo Mai itself where would-be terrorists and "revolutionaries" would receive basic training in weapons, with instructions given by ex-Legionnaires and professional terrorists. Thus, the "environmentalist" communalism of Longo Mai, made famous by the "generation of 68" was only a cover for terrorist 'basic training' conditioning and preparation.

"The next stage is intake into a terrorist group, with further training at another camp anywhere in the Mediterranean region, including Israel. It should be noted that this intake-process for screening of potential terrorist recruits is identical with the development of the hard-core of the environmentalist movement. The terrorists are distilled from the environmentalists, Maoist movements or recruited from among prison inmate, largely through either in-prison 'therapy' programs or through 'half-way house' programs. A convict with a pre-established propensity for crimes of violence and a potential psychosis is the most easily deployed, expendable sort of recruit to terrorist and related operations.

"Neither the terrorist nor the typical terrorist support group in any way corresponds to the 'myth of the idealist' misled into excesses in one way or another. But, apart from the fully conscious controllers among them, are criminalized psychotics or semipsychotics, either selected because of those preexisting qualities, or because various phases of conditioning bring about these states."[51]

According to Tavistock profiling of the terrorists, "the individual is able to perform functions involving his immediate person and environment. He can function. Where his psychosis shows most clearly is in his inability to cope sanely with abstractions concerning the world in general. In other words, he lives at best within a very small world, a world defined by a very short span of social distance surrounding his or her person."[52] Tavistock controllers then delve deeper, giving a remarkable insight into the control mechanisms of the subject. He functions within a scope of a "surrogate family," more or less as a disturbed but functionally effective two-to-four year old might – to be more exact, an adolescent or adult parody of such a two-to-four year old condition. The world outside his or her "family" is not real. His or her outer world behavior is biased toward the schizophrenic-symbolic, rather than the real. His sense of that world of symbols is magical, not cause-and-effect. He or she carries the infantile superstition of the believer in astrology toward an extreme.

Now, if such a person once agreed that "bad" people ought to die, "so they can't hurt good people," and if a "person is also dedicated to killing the 'bad' people himself, it suffices – to make the point short – to insert a face, a name and so forth in place of the symbol 'bad', and say 'we must kill that bad person', and there you have the gist of it."[53]

That's where the controlled media comes into play. The key to aiding these zombied terrorists' deployment into action are certain "code phrases." For example, the code phrase used in organizing the Baader-Meinhof assassination of Dresden Bank's Jurgen Ponto in 1977 was the epithet "Nazi Communist."

How would such a label be used by the controllers to motivate a would-be terrorist to kill his intended target? "If one knows the doctrine of Winston Churchill, Hugh Trevor-Roper, and John Wheeler-Bennett, the doctrine they set forth in defending their preference for Adolf Hitler to the insurgent German generals of 1938 and during the war, one knows that the British not only have a special hatred for German industrialists and the old Catholic Center Party of Germany, but that the British, who created Hitler, saw fit to accomplish two objectives with one argument, the argument that Nazism was a product of German industry.

"Although the myth that German industrialism created Hitler is in wide circulation, the myth has strong primary emotional significance only in two primary circles: British intelligence circles and a certain section of Zionist circles. In Zionist circles it serves as a reaction-formation in respect to Warburg backing of Hjalmar Schacht, and of Schacht's placing Hitler in the Chancellery, and of various Rothschild, Oppenheimer and other involvement in supporting Hitler before and after 1933 into the 1936-1938 period.

"This emotionally laden identification of German industrialism as being axiomatically 'Nazism' spills over from British intelligence and certain Zionist circles into elements of the anarchist-environmentalist ultra-left, especially to those layers surrounding and including 'left' terrorist gangs . In these circles, 'Nazi' means essentially 'German industrialism', and is extendable to any pro-industrial, pro-technological-progress personality of influence or political current which is viewed as 'practically the same thing as German industrialist.'"[54]

To leave no confusion concerning the significance of Zionism in this case, an interpolated clarification respecting the command-structure coordinating international terrorist operations should be made now. The key elements controlling international terrorism include the "fascist international," coordinated by the "Eastern European branch of the Maltese Order, the Sovereign Order of the Knights of St. John of Jerusalem. They also include British Round Table and official intelligence organizations. They also include elements of the Socialist International and elements of international Zionism."[55]

CREATING THE COUNTERCULTURE

The undeclared cultural war against America's youth began in earnest in 1967, when Tavistock began using open-air rock concerts to attract over four-million young people to the so-called "festivals." Unbeknownst, the youth became the victims of planned, wide-scale drug experimentation.[56] Hallucinogenic drugs such as dimethoxyphenylethylamine (STP), phencyclidine (PCP), dextromethorphan, methylenedioxyamphetamine (MDA), dimethyltryptamine (DMT), and the Beatles-promoted LSD[57] and Blue Micro Dot LSD-25 were freely distributed at these concerts. Before long, over 50 million of these attendees between the ages 10 to 25 years old returned home to become the messengers and promoters of the new drug culture, or what came to be called the "New Age."

Hallucinogenic drugs are psychomimetic, meaning that they mimic certain aspects of psychosis. Through the use of hallucinogenic drugs, one can induce temporary symptoms of psychosis and schizophrenia. Most users of hallucinogenic drugs at the time experienced whole personality changes causing total alterations of the senses.[58]

The intention of the LSD drug scene and the controlled environment it represented wasn't accidental, but completely intentional. The Tavistock Institute had extensively studied the relationship between brain and behaviour caused by hallucinogenic drugs. "Later on, the knowledge gleaned from the research was channelled into marketing MTV and radio stations through 'classic oldies' songs from 15 to 25 years ago, which are targeted at the adult population."[59]

Several Tavistock studies showed that a song or piece of music associated with one's childhood, when heard later in life, could call forth memories and associations of that earlier period. Encoded memories of popular music in the listener were recalled when he or she heard the same piece of music. These memories triggered an emotional drug flashback that set off an infantile emotional state that brought the listener back to that time, in which he or she experienced either an identical or one mirroring the drug reaction itself.[60]

Monterey Pop

The first commercial American rock festival, officially dubbed "The First Annual Monterey International Pop Festival," was held in June 1967, two years before Woodstock, when over 200,000 young people gathered in and around the Monterey County Fairgrounds in Northern California for a three-day celebration. Monterey Pop 1967 was the general rehearsal for the widespread distribution of a new type of drugs, classified as psychedelics or hallucinogens, such as LSD, which were to become prominent and wide-spread at later outdoor festivals like Woodstock.

Author Robert Santelli, in his book, *Aquarius Rising*, writes: "LSD was in abundance at Monterey. Tabs of 'Monterey Purple' [LSD-type substance also called Purple Haze] were literally given to anyone wishing to experiment a little." The two characters responsible for its distribution at Monterey Pop were a mystery man from Coco Beach Florida called Peter Goodrich and the legendary CIA contract agent code-named "Coyote."

One of the organizers of the Monterey festival was John Phillips, a member of the Mamas and the Papas rock group[61] and former press agent for the Beatles. Phillips was closely linked with Roman Polanski and his wife Sharon Tate, Mama Cass [lead singer of the Mamas and the Papas], Dennis Wilson of the Beach Boys and many other Hollywood celebrities who in turn were linked with Charles Manson's Family[82]. Manson Family members stabbed to death an eight-months-pregnant actress Sharon Tate on August 9, 1969 at her home in Los Angeles.

Mama Cass and John Phillips were connected to Manson through "The Process Church of the Final Judgment," an offshoot of the Church of Scientology founded in England in the mid-60s by a couple of former Scientologists, Robert and Mary Ann DeGrimston. Formed sometime in 1963-64, PCFJ was a mixture of reincarnation, existentialism, an attempt to merge the worships of Jehova and Lucifer, and a bit of neo-Nazi flavour. It is significant that "The Process's" legal work was handled by an elite Wall Street law firm, Morris and McVeigh, whose main backer was the American Family Foundation, who used intelligence-connected 'mind-control' experts such as LSD researcher Dr. Louis Jolyon West to orchestrate 'cult-anti-cult' hysteria. Dr. West was a major participant in the CIA's MK-ULTRA, which came out of the Tavistock Institute's studies of Nazi social control techniques.

People with only a passing knowledge of the Manson case usually don't realize that "the Beach Boys recorded one of Manson's songs and released it as the 'B' side of their cover of the old Ersel Hickey 1958 hit "Bluebirds Over the Mountain'" a song about lost love."[62] The title of the song was suggestive of the CIA mind-control operation. Bluebird, was a program for exploring the uses of hypnosis and other means to protect the recently created CIA and its personnel from enemy psychic penetration – BLUEBIRD. Is there a reason Manson's song was on the flipside of the "Bluebird" cover? Had the CIA's "bluebird" escaped? Was Manson a "bluebird," i.e., a mind-control subject, gone amok?

As Peter Levenda explains in *Sinister Forces*, "The 'land of memory,' was the target of the Bluebird project: to enter that 'land' in another person's mind, to go through the drawers, re-aranging the furniture, and leave unnoticed. Korean War was fought to a stalemate."[63] Vietnam was raging all around us. Africa was ablaze with 'godless communists' holding the upper hand. In America, the year 1968 was the pivotal year in the lives of most people of the generation. It was a time of multiple assassinations, civil unrest, the riots at the Democratic National Convention in Chicago, and the election of Richard Nixon as President. For many, it signalled the death of a dream. For some the rebirth of a nightmare. It was the year when those things that are now so secret, so classified and behind the scenes were up front and in your face: the Cold War, the CIA, Prague Spring.

It was the year that Frankfurt School, Tavistock Institute, Institute for Policy Studies and other like-minded organizations went to war against America. It was the year of Woodstock and the Age of Aquarius and the death of all that we held dear. It wasn't only "flowers in their hair" that they got; they also got scrambled brains and total dependency on narcotics, a shortened lifespan, a degraded life-style which tended to sink lower and lower into the mire.

THE AGE OF AQUARIUS: WOODSTOCK MUSIC FESTIVAL

The largest concert of all time, the open-air 'Woodstock Music and Art Fair,'[64] was what *Time* magazine labelled an 'Aquarian Festival' and 'history's largest happening.' In the process, Woodstock became part of the cultural lexicon. Publicists carefully chose the term, 'Aquarian.'

According to astronomers, the Ages progress in retrograde motion, going in the opposite direction as the sun, which moves from the Age of Aquarius into the Age of Pisces into the Age of Aries, and so forth. If you follow astrological beliefs, the Age of Pisces, (an artefact of the Precession of the Equinoxes) is the time span from about 200 B.C. until the current day. Approximately every 2,160 years, the Precession of the Equinoxes appears to rotate the spring equinox from one constellation to another. We appear to be ending the Age of Pisces and beginning the Age of Aquarius. Moving into the Aquarian age signifies that the Age of Pisces, which is the age of Christ, has come to an end.[65]

"At Woodstock," writes journalist Donald Phau, "nearly half a million youth gathered to be drugged and brainwashed on a farm. The victims were isolated, immersed in filth, pumped with psychedelic drugs, and kept awake continuously for three straight days, and all with the full complicity of the FBI and government officials."

Once again, it was the networks of British Military Intelligence and the CIA, who were the initiators.

A man credited with Woodstock's creation was Artie Kornfeld, the director of Capitol Records, which is owned by EMI. EMI, led by aristocrat Sir Joseph Lockwood, stands for Electrical Music Industries, and unknown to most, is one of Britain's largest producers of military electronics and a key member of Britain's military intelligence establishment as a military contractor to the

British War Office. Furthermore, EMI's silent partner is another major record label, RCA, which is also active in military and space electronics and satellite communications. It is a classic example of what President Dwight Eisenhower would have called the "military industrial complex," but is really the One World Government solidifying everything into fewer and fewer hands.[66]

The money for Woodstock was supplied by John Roberts, the heir of the large Pennsylvania-based Block Drugs pharmaceutical empire and toothpaste manufacturing fortune. Joel Rosenmann, one of the three partners, writes: "As the concert neared, food and water were clearly going to be in short supply, sanitary facilities overtaxed, tempers short, drugs overabundant. Worst of all, there was no way for anyone who wanted to, to leave."[67]

According to John Roberts, sitting in your own excrement was part of the original plan. "We're going to hand out bananas at the gate to bind our patrons."

Security for the concert was provided by a hippie commune called the Hog Farm trained in the mass distribution of LSD. The Hog Farm's connection to drug distribution, however, was well known to concert organizers. John Roberts, Woodstock founder and director writes: "Their fee was simply transportation to and from the festival ... a peace-keeping force that looked, talked, and smelled like the crowd would be both highly credible and highly effective... and most important, they were wise in the ways of drugs, knowing good acid from bad, good trips from bummers, good medicine from poison, etc."[68]

But, the famed undercover hippie commune, Hog Farm, was none other than the West Coast psychedelic group called the "Merry Pranksters," whose leader was none other than Ken Kesey – the same Kesey who wove his experiences as a volunteer in medical experiments into his book, *One Flew over the Cockoo's Nest*.

THE EXPERIMENT BEGINS

Donald Phau, in his book the *Satanic Roots of Rock*, describes the event: "Two days before the scheduled start of the concert, 50,000 kids had already arrived in Woodstock. Drugs immediately began to circulate." Phau, in the same book, continues: "For the next three days, the nearly half a million young people that arrived were subjected to continual drugs and rock music. Because of torrential rains, they were forced to wallow in knee-deep mud. There were no shelters, and no way to get out. Cars were parked over eight miles away.z

Rosenmann writes that the key to the 'Woodstock experiment' was "keeping our performers performing around the clock...to keep the kids transfixed..." On August 17, the *New York Times* reported: "Tonight, a festival announcer warned from the stage, that 'badly manufactured acid' [a term for LSD] was being circulated. He said: 'You aren't taking poison acid. The acid's not poison. It's just badly manufactured acid. You are not going to die.... So if you think you've taken poison, you haven't. But if you're worried, just take half a tablet.'" The advice, to nearly 500,000 people, "just take half a tablet" was given by none other than MK-ULTRA agent Wavy Gravy."

It would be another decade before the counterculture became part of the American lexicon. But the first seeds of what was a giant, secret project to turn the values of America inside out were sown. Drugs, sex, rock and roll, massive nation-wide demonstrations, hippies, communes, drop-out drug heads, the Nixon presidency, the Vietnam War were tearing the fiber of American society apart. The old world and the new collided head on, unbeknownst to the "flower children" that it was all part of a secret, social and occult agenda[69] designed by some of the most brilliant and diabolical minds in the world. Hidden away in prestigious foundations, corporations and think-tanks, these political and financial puppeteers ruthlessly and cunningly used pop culture to foster alienation and dysfunction to achieve arrested development and family breakdown for personal gain.

Through 1965-1969, the psychedelic movement gained much momentum and culminated with events like Woodstock. However, quick as it came, it was gone. LSD was outlawed. Jimi Hendrix and Janis Joplin died. Timothy Leary dropped off his soap-box, and the United States, after failing miserably in Vietnam, drifted into a depressing 1970s. Yet once again, Tavistock was waiting, ready to exploit that apathy for their personal gain.

THE AQUARIAN CONSPIRACY

In the spring of 1980, a book exploded on the scene called *The Aquarian Conspiracy*. It sold over one million copies and was translated into 10 languages. Overnight, it became a manifesto of the counterculture. Defining the counterculture as a cultural group whose values and norms are at odds with those of the social mainstream, and in practical terms as the youth rebellion that swept North America and Western Europe in the 1960s and early 1970s, *The Aquarian Conspiracy* declared that it is "now time for the 15 million Americans involved in the counterculture to join in bringing about a radical change in the United States."[70]

In fact, this book was the first mass publication to promote teamwork, a concept hailed as a paragon of virtue and readily promoted by business management "gurus." Teamwork, however, was a psychological technique the Tavistock Institute introduced as a way to control brilliant individuals through peer pressure. It was part of the experiment conducted by Brigadier-General and Tavistock master John Rees in the field of group therapy. L. Marcus, writing for *The Campaigner* magazine in April 1974, explains: "A skilled group leader can use the group to create a powerful 'family' environment. Once this environment is induced, it becomes possible for the therapist to manipulate a member of the group, not by a direct attack, but by subtly manipulating through suggestions the other members of the group. If the victim has been sucked into thinking of the group as something warm and helpful, then, when that environment has been manipulated to turn against him, it will tend to have the impact of deep motherly rejection. Furthermore, if the victim is not completely aware of the therapist's chain of manipulation, he will tend to internalize their result, thinking that he himself is responsible for this new awareness about himself."

Marcus goes on to say: "...If the controllers can succeed in structuring a stressed individual's or group's situation appropriately, the victim(s) can be induced to develop for himself a special sort of 'reaction formation' through which he 'democratically' arrives precisely at the attitudes and decisions the dictators wish to force upon him."

So, how did author Marilyn Ferguson see the entire Aquarian movement? "While outlining a not-yet-titled book about the emerging social alternatives, I thought again about the peculiar form of this movement: its atypical leadership, the patient intensity of its adherents, their unlikely successes. It suddenly struck me that in their sharing of strategies, their linkage, and their recognition of each other by subtle signals, the participants were not merely cooperating with one another. They were in collusion. It – this movement – is a conspiracy!"[71]

In fact, the counterculture revolution was a cleverly concealed conspiracy from the top down, created as a method of social control through the use of difficult-to-identify, silent-weapons technology and "used to drain the United States of its commitment to scientific and technological progress."[72]

Initially, *The Aquarian Conspiracy* was a secret U.S. government project assigned to Stanford Research Institute(SRI) and created in 1946 by the Tavistock Institute to study how newly created cultural and social trends affect society such as libertarianism, left-libertarianism, liberalism, socialism, anarchism, communism, materialism, naturism, mysticism, hedonism, spirituality, environmentalism, feminism, New Age and many other basic orientations on society.

Never publicly released and secretly named, *The Changing Images of Man*,[73] *The Aquarian Conspiracy* sprang from a Policy Report prepared by the Stanford Research Institute Center for the study of Social Policy.[74] The 319 page mimeographed report was prepared by a team of fourteen researchers and directed by a panel of 23 controllers including Margaret Mead, B.F.Skinner, Ervin Laszlo of the United Nations, Sir Geoffrey Vickers of British Intelligence. The entire project was overseen by Professor Willis Harmon, a "futurist" whose speciality was promoting a post-industrial social paradigm as a popular version of how to transform or "soften up," according to Tavistock manuals, the United States into Aldous Huxley's *Brave New World.*

In a 1961 lecture, Aldous Huxley had described this police state as "the final revolution" – a "dictatorship without tears," where people "love their servitude." He said the goal was to produce "a kind of painless concentration camp for entire societies so that people will, in fact, have their liberties taken away ... but ... will be distracted from any desire to rebel by propaganda or brainwashing ... enhanced by pharmacological methods."

The aim of the Stanford study, state the authors, is to change the desire of mankind from that of industrial progress to one that embraces *"spiritualism."* [War by non-violent or non-linear confrontation] "The study asserts that, in our present society, the 'image of industrial and technological man' is obsolete and must be 'discarded.'

"Many of our present images appear to have become dangerously obsolete, however ... science, technology, and economics have made possible really significant strides toward achieving such basic human goals as physical safety and security, material comfort and better health. But many of these successes have brought with them problems of being too successful – problems that seem insoluble within the set of societal value-premises that led to their emergence ... Our highly developed system of technology leads to higher vulnerability and breakdowns. Indeed the range and interconnected impact of societal problems that are now emerging pose a serious threat to our civilization ... If our predictions of the future prove correct, we can expect the association problems of the trend to become more serious, more universal and to occur more rapidly."[75]

Therefore, the SRI study concludes, "we must change the industrial-technological image of man fast: Analysis of the nature of contemporary societal problems leads to the conclusion that . . . the images of man that dominated the last two centuries will be inadequate for the post-industrial era."[76]

The report concludes with the following set of predictions: "There are numerous indications that a new, anticipatory image of mankind may be emerging.

1. Youth involvement in political processes
2. Women's liberation movement
3. Youth rebellion against societal wrongs
4. Emerging interest in social responsibility of business
5. The generation gap implying a changing paradigm
6. The anti-technological bias of many young people
7. Experimentation with new family structures and interpersonal relationships
8. The emergence of communes as alternative lifestyle
9. The emergence of conservative/ecology movement
10. A surge in interest in Eastern religious and philosophical perspectives
11. A renewed interest in 'Fundamentalist' Christianity
12. Labor union concerns with quality of the work environment
13. An increasing interest in meditation and other spiritual disciplines
14. The increasing importance of 'self-realization' process

These disparate trends do not, when taken individually, signify the emergence of a new image of a human being; yet when they are considered collectively, they suggest substantial societal stirrings which may eventually emerge into a new and guiding image."[77]

This was published in 1974. In retrospect, how many of these hideous, inhuman forms have become accepted by the mainstream? Each and every one of them. Investigative journalist Criton Zoakos called it "countercultural zombiizm."[78]

"Thus softened up," the United States "was now deemed ripe for the introduction of drugs [especially cocaine, crack cocaine, heroin] to rival the prohibition era in scope and the huge amounts of money to be made. This too was an integral part of the Aquarian Conspiracy. The proliferation of drug usage was one of the subjects under study at the Science Policy Research Unit [SPRU] at the Tavistock Institute's Sussex University facility. It was known as the 'future shocks' center, a title given to so-called future oriented psychology designed to manipulate whole population groups to induce 'future shocks.'"[79]

The Columbus Center at the University of Sussex is a Tavistock-front organization. The Center is involved only in one line of work: it publishes 'Studies in the Dynamics of Persecution and Extermination.' In other words, "the United States were to be programmed to change and become so accustomed to such planned changes that it would hardly be noticeable when profound changes did occur."[80]

In fact, this far-reaching initiative for silent-weapon technology was first discussed and put into action as an official doctrine by the Policy Committee of the Bilderberg Group at their inaugural meeting in 1954. Then they used the term 'Quiet War' [WWIII] to describe the overt tactical methodology to subjugate the Human race. The document, titled *TOP SECRET: Silent Weapons for Quiet Wars, An introductory Programming Manual*, was uncovered quite by accident on July 7th, 1986, when an employee of the Boeing Aircraft Co. purchased a surplus IBM copier for scrap parts at a sale and discovered inside details of a plan, hatched in the embryonic days of the 'Cold War.' This manual of strategies called for control of the masses through manipulation of industry, peoples' pastimes, education and political leanings. It called for a quiet revolution, pitting brother against brother, to divert the public's attention from what is really going on.

Here is a partial quote from this document (TM-SW7905.1): "It is patently impossible to discuss social engineering or the automation of a society, i.e., the engineering of social automation systems (silent weapons) on a national or worldwide scale without implying extensive objectives of social control and destruction of human life, i.e., slavery and genocide. This manual is in itself an analog declaration of intent. Such writing must be secured from public scrutiny. Otherwise, it might be recognized as a technically formal declaration of domestic war. Furthermore, whenever any person or group of persons in a position of great power and without full knowledge and consent of the public, uses such knowledge and methodology for economic conquest – it must be understood that a state of domestic warfare exists between said person or group of persons and the public. The solution of today's problems requires an approach, which is ruthlessly candid, with no agonizing over religious, moral or cultural values. You have qualified for this project because of your ability to look at human society with cold objectivity, and yet analyze and discuss your observations and conclusions with others of similar intellectual capacity without a loss of discretion or humility. Such virtues are exercised in your own best interest. Do not deviate from them."

As a side note, substantial chunks of the 3,000 pages of "recommendations" given to the newly-elected Ronald Reagan administration in January 1981 were based upon material taken from Willis Harmon's *Changing Images Of Man*.

During the full moon, on December 8, 1980, John Lennon was killed by a man named Mark Chapman. A man who prayed to Satan hours before he pulled the trigger, who walked robotically through the motions of murder, in a trance-like daze, and found himself in police custody, staring bewilderedly out at the world from within a tiny closet of darkened dreams. A carefully constructed multiple personality with a tightly focused purpose, a single target, and ultimate deniability, who stood and waited for police to arrive, calmly reading *The Catcher in The Rye*, that coming-of-age novel by former US Army intelligence officer J.D. Salinger. It is doubtful we shall ever learn whether Mark Chapman was definitively the victim of an artificially induced *model psychosis*, "Manchurian Candidate" type assassin deployed by the combined Tavistock/CIA/MI6 to silence the increasingly unmanageable Lennon. In the end, it is the dreamers who die by an assassin's hand. Quite often, they are killed by men who themselves are dreamers.

Eight months later on August 1, 1981, a new concept was created in television that was part of the Tavistock Institute-led Quiet War for silent weapons technology. It was a 24-hour per-day music station called MTV, Music Television. Popular fairy tales aside, MTV, too, was created by the controllers at the world's leading think-tanks and brainwashing institutes for the purpose of further destroying the youth culture and permanently implanting another "generational shift" in modern society.

Endnotes

1. The Frankfurt School and Political Correctness, Michael Minnicino, *Fidelio Magazine*, winter 1992

2. *Mind Control World Control*, Jim Keith, Adventures Unlimited Press, 1977, ISBN: 978-0932813459

3. The Frankfurt School and Political Correctness, Michael Minnicino, *Fidelio Magazine*, winter 1992

4. *Understanding Understanding*, Dr. Humphry Osmond, John A. Osmundsen, Jerome Agel, Harper & Row, 1974, ISBN: 978-0060132392

5. Clients, include the Pentagon, AT&T, Chase Manhattan Bank, IBM, Republican Party, U.S. Air Force, U.S. Department of Energy and NASA. The interlocking leadership between the trustees at Rand, and the Ford, Rockefeller, and Carnegie foundations is a classic case of CFR/ Bilderberg modus operandi. Ford foundation gave one million dollars to Rand in 1952, at a time when the president of Ford Foundation was simultaneously the chairman of Rand.

6. *Foundations: Their Power and Influence*, Rene Wormser, p65-66, Sevierville TN: Covenant House Books, 1993

7. According to RAND paper, document number P-2575, re-edition 2004, "an experiment that would attempt to measure any long-lasting changes in attitudes, values, and communicative ability resulting from the administration of LSD. In particular, the measures would concentrate on changes in closed-mindedness as reflected by scales of dogmatism, opinionation, and ethnocentricity."

8. The Frankfurt School and Political Correctness, Michael Minnicino, *Fidelio Magazine*, winter 1992

9. From Cybernetics to Littleton: Techniques of Mind Control, Jeffrey Steinberg, *EIR*, April 2000

10. Ibid

11. Ibid

12. Ibid

13. *Understanding Understanding*, Dr. Humphry Osmond, John A. Osmundsen, Jerome Agel, Harper & Row, 1974, ISBN: 978-0060132392

14. *Sinister Forces*, Peter Levenda, Trineday Press, 2011, p.218-219

15. *Star Wars and Littleton*, Lyndon H. LaRouche, Jr., June 11, 1999

16. Ibid

17. From Cybernetics to Littleton: Techniques of Mind Control, Jeffrey Steinberg, *EIR*, April 2000

18. How to Profile the Terrorist, *EIR*, Volume 5, number 37, September 26, 1978

19. Ibid

20. *The House of Life: Magic and Medica' Science in Ancient Egypt*, Paul Ghalioungui, New York: Schram Enterprises, 1974.

21. *The Impact of Science on Society*, Bertrand Russell, reprint edition, Routledge, 1985, ISBN: 978-0415109062

22. The 'No-Soul' Gang Behind Reverend Moon's Gnostic Sex Cult, Larry Hecht, *EIR*, December 20, 2002

23. Ibid

24. Huxley, Isis, LSD and the roots of the American hedonist culture, *EIR*, 1980

25. The Aquarian Conspiracy, *Executive Intelligence Review*, Volume 7, Number 18, May 13, 1980, p.21

26. H.G. Wells, *Anticipations of the Reaction of Mechanical and Scientific Progress Upon Human Life and Thought*, New York: Harper and Row, 1902, p.285.

27. Huxley, Isis, LSD and the roots of the American hedonist culture, *EIR*, 1980

28. Most readers naturally side with the alienated feelings of John in *Brave New World* and associate the multi-world religious views that he practices as some sort of rebellion – a spiritual escape from the evil clutches of the "Utopian" society – when in reality the New World Order system, which Huxley and his family and friends all worked for, will actually foster a multi-faceted one world religion as the corner-stone to help bind all nations under such a one world government.

29. Larry Hecht, H.G. Wells and Bertrand Russell: The 'No-Soul' gang Behind Reverend Moon's Gnostic Sex Cult, *EIR*.

30. *Isis Unveiled, A Master Key to the Mysteries of Ancient and Modern Science and Theology*, Helena P. Blavatsky, Los Angeles: Theosophy Co., 1931.

31. The Aquarian Conspiracy, *Executive Intelligence Review*, Volume 7, Number 18, May 13, 1980, p.21

32. From Cybernetics to Littleton: Techniques of Mind Control, Jeffrey Steinberg, *EIR*, April 2000

33. Jim Morrison married his wife at a Satanic Wicca wedding standing in a pentagram and drinking each others blood. The back cover of the Doors album "13" shows the group gathered around a bust of Aliester Crowley.

34. Bankers behind counter culture, Henry Makow, www.savethemales.ca, May 7, 2005

35. *Steps to the Ecology of the Mind*, Gregory Bateson, New York: Chandler, 1972

36. The Tavistock Grin, Michael Minnicino, *The Campaigner*, April 1974

37. In 1967, Tavistock sponsored a Conference on the "Dialectics of Liberation," chaired by Tavistock psychoanalyst Dr. R.D. Laing, himself a popularized author and advocate of drug use.

38. The Aquarian Conspiracy, *Executive Intelligence Review*, Volume 7, Number 18, May 13, 1980, p.21

39. The Sequoia Seminars – 1954 – LSD Therapy – History http://forum.prisonplanet.com/index.php?topic=153737.30;wap2

40. *Stamp Out the Aquarian Conspiracy*, Criton Zoakos, Citizens for LaRouche monograph, New York, 1980, pp. 60-63.

41. Call to Investigate the Institute for Policy Studies, *EIR*, Volume 4, number 26, June 28, 1977, p.1

42. The War Of Ideas In Contemporary International Relations, Georgii Arbatov Theories And Critical Studies Series Progress Publishers, Moscow 1973 English Language edition 317 pages, hard-cover , reviewed in *EIR* Volume 4, Number 19, May 10, 1977

43. The Real CIA – the Rockefellers' Fascist Establishment, L. Marcus, *The Campaigner* magazine, April 1974

44. Call to Investigate the Institute for Policy Studies, *EIR*, Volume 4, number 26, June 28, 1977, p1

45. Ibid

46. Ibid

47. Ibid

48. Ibid

49. How to profile the terrorist infrastructure, *EIR*, Sept 26-oct 2, 1978, p.44

50. Ibid

51. Ibid

52. Ibid

53. Ibid

54. Ibid

55. Ibid

56. In June 1969, 150,000 people turned out for the Newport Pop Festival at Devonshire Downs in Northridge, California. One week later another 50,000 fans showed up in Mile High Stadium in Denver, Colorado for the Denver Pop Festival. Both of those 3-day events were seriously marred by violence and drugs as thousands of ticketless gate-crashers battled with cops. FBI officers on site admitted to being given orders not to interfere with drug pushers.

57. During this time, John Lennon was becoming increasingly aware of continued manipulation of thought on the part of those who held the power. Lennon was aware that the mind control drug LSD in the CIA/Tavistock/MI6 arsenal, was having adverse effect on the population, to which it had been funneled so extensively. Instead of controlling the people, LSD liberated them. *Playboy* interview, 1975, John Lennon and Yoko Ono

58. *Los Secretos del Club Bilderberg*, Daniel Estulin, editorial Planeta, Septiembre 2006

59. Ibid

60. Turn Off Your TV, Part 12 – MTV Is the Church of Satan, L. Wolfe, *The New Federalist*

61. On June 10, 1967, one week before the First Monterey Pop festival, Phillips released a song called "San Francisco," which sold over five million copies. The song called for youth [new "hippie" movement] throughout the country to "come to San Francisco with flowers in their hair." Some who came became the victims of Charles Manson and his Family, who recruited his cult- "family" exclusively from runaway youth.

62. *Sinister Forces – The Manson Secret: A Grimoire of American Political Witchcraft*, Peter Levenda, TrineDay Press, 2001

63. Ibid

64. Actually, the Woodstock Music festival took place 110 kms west of Woodstock, New York State, in an alfalfa pasture carved out of rolling Maplewood forests.

65. Anti-Christ cultist Nietzsche announced that the twentieth century would see the end of the Age of Pisces, which Aquarians associate with the figures of Socrates and Christ; Satanist Friedrich Nietzsche prophesied that the New Age would be the Age of Aquarius, which he identified with the Satanic figure Dionysos.

66. *Los Secretos del Club Bilderberg*, Daniel Estulin, editorial Planeta, Septiembre 2006

67. The Satanic Roots of Rock, Donald Phau, the konformist.com

68. Ibid

69. Over the past 50 years, research in the fields of psychology, sociology and psychiatry has shown that "there are clearly marked limits to the amount of changes and the nature of them that the mind can deal with." According to Science Policy Research Unit [SPRU] at Tavistock's Sussex University facility, 'future shocks' is defined "as physical and psychological distress arising from the excess load on the decision-making mechanism of the human mind." In other words, "a series of events, which come so fast that the human brain cannot absorb the information." After continuous shocks, the large targeted population group discovers that it does not want to make choices any more. Apathy takes over, often preceded by mindless violence such as is characteristic of the Los Angeles street gangs in the 1960s and the 1980s. "Such a group becomes easy to control and will docilely follow orders without rebelling, which is the object of the exercise."

70. The Aquarian Conspiracy, *Executive Intelligence Review*, Volume 7, Number 18, May 13, 1980, p.21

71. Marilyn Ferguson, *The Aquarian Conspiracy*, Los Angeles: J.P. Archer, 1980, p.19.

72. Chapter VII of EIR, *DOPE, INC*, Lyndon LaRouche, 3rd Ed. 1992

73. May 1974, Contract Number URH (489)-2150-Policy Research Report No. 414.74

74. SRI was also deeply involved in the notorious MK-ULTRA program. Harman was a long time president of the Institute of Noetic Sciences and a friend to Edgar Mitchell, also of IONS, who in turn, was a long time friend of George Bush Sr. (Both are 33rd degree Scottish rite masons).

75. The Aquarian Conspiracy, *Executive Intelligence Review*, Volume 7, Number 18, May 13, 1980, p.21

76. Ibid

77. Contract number URH 489-2150, Policy Research Report #4/4.74

78. Criton Zoakos, The Aquarian Conspiracy's road to Orwell's 1984, *EIR*, Volume 7, number 18, May 13, 1980, p.27

79. John Coleman, *Conspirators' Hierarchy: The Story of the Committee of 300*, Global Review Publications, 2006

80. The report, one of the most far-reaching investigations into how man might be changed, covering 319 pages, was written by "14 new science scientists under the supervision of Tavistock and 23 top controllers including psychologist B. F. Skinner, [American experimental psychologist, member of Tavistock and an advocate of behaviourism as a function of environmental histories of reinforcement as well as an author of controversial works in which he proposed the widespread use of psychological behaviour modification techniques as a means of human control], anthropologist Margaret Mead, Ervin Lazlo of the United Nations and Sir Geoffrey Vickers, a high-level British intelligence officer in MI6

CHAPTER FIVE

Television

T he slylest form of control is making people think they are free while having someone manipulate their every move. One form of dictatorship is being in a prison cell, seeing the bars. The other, a far more subtle form of control and dictatorship is when you can't see the bars and you think you are free. The biggest hypnotist in the world is an oblong box in the corner of the room that tells people what to believe. Television, with its reach into everyone's home, creates the basis for the mass brainwashing of citizens, as we will see in this chapter. You may not realize it, yet, but your mind is being shaped and moulded every time you turn the one-eyed baby sitter on.

When you watch TV, the right hemisphere is twice as active as the left, which in itself is a neurological anomaly. "The crossover from left to right releases a surge of the body's natural opiates: endorphins, which include beta-endorphins and enkephalins. Endorphins are *structurally identical* to opium and its derivatives (morphine, codeine, heroin, etc.)."[1] In other words, your television works as a high-tech drug delivery system, and we all feel its effects. Another effect of watching television is that the "higher brain regions such as the midbrain and the neo-cortex, are shut down, and most activity shifts to the limbic system, your lower brain region. The lower or reptile brain simply stands poised to react to the environment using deeply embedded 'fight or flight' response programs. Moreover, these lower brain regions cannot distinguish reality from fabricated images (a job performed by the neo-cortex), so they react to television content as though it were real, releasing appropriate hormones and so on. Studies have proven that, in the long run, too much activity in the lower brain leads to atrophy in the higher brain regions."[2] Now, you may not have known this, but I bet you the people who brainwash you on daily basis, surely did. Don´t worry, we will come back to this a bit later.

For the brainwashers in charge of this societal transformation, they have pulled off the ultimate trick. They were able to persuade people that what they can see with their eyes is what's there to see. Subsequently, people will laugh in your face when you try to explain to them the bigger picture and the unseen reality behind the curtain.

In a 1981 interview, Hal Becker from a think-tank called "Futures Group" in Connecticut said: "I know the secret of making the average American believe anything I want him to. Just let me control television.... You put something on the television and it becomes reality. If the world outside the TV set contradicts the images, people start trying to change the world to make it like the TV set images...."

You see, television is not the truth, for as much as people keep tuning in to its lies. Television is an amusement park, a group of jugglers, belly dancers, storytellers, singers and stripers. But people, have been completely hypnotised by the Tube. You sit there, day after day, night after night.... TV is all most of you know! Five percent of Americans read more than five books per year, yet one billion people tune in to the Oscar awards. You dream like a tube, speak like a tube, smell, dress, act like your television. Most people feel they are better acquainted with Paris Hilton, Britney Spears or Lady Gaga than with their own husband or wife. This is madness! Don't you see? How many millions of you are prepared to believe anything the tube tells you? What's more, huge numbers of people at the top are prepared to tell you anything in the name of 'war against terror', audience share and advertising dollars as long as you vote for them, buy their products and allow them to brainwash you. "Television provided an ideal means to create an homogenous culture, a mass culture, through which popular opinion could be shaped and controlled, so that everyone in the country would think the same."[3]

Why is it being done? To dumb you down. To brainwash you. To turn you into a touchy-feely adult with infantile tendencies. So you don't get in the way of important people by doing too much thinking on your own. Think about it! Over 75% of people get all their information from the television. In fact, the only truth most people know is what they get over the television. There is an entire generation of people who did not know anything that didn't come out of this tube. This tube has become the Gospel, the ultimate revelation. It can make or break presidents and Prime ministers. This tube is the most awesome force in this God damn world. However, what would happen if it got into the hands of the wrong people? And when the largest company in the world controls the most awesome propaganda force in the entire Universe, who knows what shit will be peddled for truth on the networks?

THE MAKING OF A FASCIST SOCIETY

Most people think they have a pretty good idea about what fascist society is all about. They have seen the pictures in the made-for-TV movies, of Nazi Germany in the 1930s. The crowds, the frenzy, the glitter, the jackbooted troops. The huge rallies, the waving of the flags. Then, there were the speeches by Hitler made to the cheering approval of enormous crowds of people, who raised their arms in salute at the beckoning of their Fuerher. Who can forget the terrifying images of Nazi thugs breaking windows in the Jewish ghetto, pulling frail old ladies by their hair onto the streets, the Gestapo and the SS troops using the butt end of their machine gun to beat someone into a pulp. Then there are other images: the death camps and the horrified world that didn´t know anything about them. The bodies, the teeth, the bones, the walking dead, and the hair and the gold teeth and the rings and the ovens.

The greatest generation of the 20th century went to war to defeat the greatest evil man has ever seen. Twenty-six million Soviets died in the war plus millions of Europeans and half a million American soldiers. Never again. That's what we all said once the war ended. Do you remember? The world would never tolerate such inhumanity again. We would never have another Hitler in our midst. Never would human rights be so flagrantly trampled on. Yes, you say. Never again! Are you sure about that? What if I told you we let the monster back in and you, reader are partially to blame? Would you believe me?

Now, take yourselves back to the victorious US troops entering Baghdad and Kabul. Think back to the war with Libya. Think back to Operation Desert Storm. Do you remember the parades, the confetti, the tickertape, the troops and the very expensive equipment parading on television in front of the adoring masses waving flags? Does that sound familiar? There were millions of people throughout the United States celebrating the great victory, waving the little plastic flags which were given out by the hundreds of thousands, the cheering approval of enormous crowds of people, who raised their arms in salute at the beckoning of their President and the commanding General. Is it coming back now? I thought it might. Yes, the thrill of victory and the agony of defeat. Except, this time, we really kicked the butts of the stinking A-rabs. Didn´t we? Hi, mom, we are number 1!

In fact, if you think back, these celebrations were organized by television. Come to think of, as was the scenario for the intervention, which was simply copied and pasted from past historical events. Think back to Sudan and the Save Darfur coalition, to Kosovo and the Balkans, to Somalia, Granada, the Falkland Islands. Television acted as a surrogate facilitator of a "largest patriotic celebration in history" as they were called.

As Keith Harmon Snow reports for *GlobalResearch* on February 7, 2007, in reference to Sudan Britain-inspired conflict. "First: Create instability and chaos that gives the appearance of Arabs fighting Africans (it's always those other people over there killing each other). Second: Wage a media campaign that focuses a laser beam of public attention on the rising instability. Third: Whip up public opinion and fury among a highly manipulated Western population who will, quite literally, believe anything. Fourth: Make sure the devil – this time it's the Janjaweed – comes on horseback. This latter point underscores the tight, unwavering narrative of good versus evil. Fifth: Demonize the "enemy" [read: dirty *A-Rabs*] and their partners [Chinese oil companies and Russian arms dealers]. Sixth: Send Christian soldiers and their "humanitarian" armies onward; Enter "Save Darfur!" and, voilà!, a movement is born. Seventh: Continue to chip away the power of the enemy by chipping away at their credibility. Eighth: Under the banners of high moral approbation, and with full support of a deeply caring Western public, overthrow the malevolent forces [of Islam and the Orient] and instil a benevolent, peace-loving, pro-democracy government. Last: Wipe away the sanctions, no longer needed, and bring much-needed 'development' to another backward country. And there you have it: yet another *civilizing* mission to conquer those barbaric Arab hoards, and those starving, helpless, uneducated, diseased, AIDS-infested, tribal Africans." Amen!

This example shows the efficiency with which television can organize public opinion on behalf of foreign military crusades that advance the cause of world government – without even explicitly stating that world government or One World Company is the goal. Remember, it was the television that told you what you were celebrating, and the why you must feel proud to be an American or at least be envious of others who are American. This doesn´t apply solely to the world of politics. Think back to Spain's 2010 "glorious" world football cup victory. The celebration wouldn't have been the same without TV reaffirming our right to feel "proud."

What you might not have realized, because you were not told this little unimportant fact, is that prior to media's "humanitarian" propaganda campaigns, neither Somalia nor Kosovo and certainly not Sudan figured prominently among the typical American's concerns. In fact, over 85% of American public could not find Sudan on the world map. The same could be said for Somalia and certainly Kosovo, not to mention Iraq prior to the Desert Storm invasion in 1991. 87% of Americans didn't know where Iraq was on the map and had no idea who Saddam Hussein was until CNN's diligent, relentless efforts to indoctrinate the American public made those military campaigns possible.

Nevertheless, what's absolutely mind-boggling is that the public never questioned any of it. Instead, people chose to participate in the celebration, either directly, lining the streets and waving US flags or through national television newscasts.

By the end of 2010, more than one and a half million innocent Iraqis are dead, along with Saddam Hussein, 5,000+ American troops and unknown tens of thousands maimed for life who are "liberating" the country on behalf of British Petroleum, Royal Dutch Shell, Halliburton, Blackwater, Chase Manhattan Bank, Bank of America, CitiGroup and an unending plethora of multinational corporations, all vying for a piece of an Iraqi sweepstakes and wealth. Hi mom!

Do you think Afghanistan is any different? Another war, another dollar, another list of unsavoury statistics. Between Afghanistan and Iraq, tenths of thousands of dead children, hundreds of thousands maimed for life. And for whom did we fight? US government overthrew the filthy Taliban, only to replace them with one of the most corrupt leaders the world has ever seen – Hamid Karzai, a major drug trafficker in bed with the same unsavoury forces we came to fight against. Hi mom! Are we still number one? Number one in stupidity is what we are. So, how do you feel now? This is what you celebrated.

And believe me, there is no difference between the jackbooted Nazi thugs from the 1930s and the jackbooted democratic liberators from 2011. Without even realizing it, you became part of a mob, a fascist mob organized entirely by television.

Still don´t believe me? Have you traveled to the United States recently? The airport personnel, part of the Homeland Security Department, are the same jackbooted Nazi goons who dragged little old Jewish ladies out of their homes and beat them to death with the rifle butt for pleasure and in the name of security. While you, the model citizen stood and stands idly by, speechless and helpless. Did it ever occur to you to step in and help? No, because you and your fellow citizens have been brainwashed by television to never question authority, no matter how shocking and anti-humane their behaviour might be. Or do you really think you became this way on your own? Think back thirty years. Did people behave this way towards their fellow men? Hardly, right? So, what happened? Along the way, somewhere, we lost our consciousness.

Consciousness is something we each develop to the degree we feel and think for ourselves. Insofar as we merely accept the emotional and intellectual commonplaces of others, we remain dead.

Let's put it this way: "The advent and mass dissemination of television technology has rendered the Nazi model for a fascist society obsolete; it has provided a better, more subtle, and more powerful means of social control than the organized terror of the Nazi state."[4]

To most people, this may sounds absolutely shocking. After all, what does an innocent celebration of democracy or football victory have to do with Nazism, right?

THE FASCIST CONCEPT OF MAN

Investigative journalist Lonnie Wolfe succinctly tells the story: "The Nazi state was created by the same oligarchical financial and political interests who today control what we call the mass media and television. Forget about whatever stories you have seen on television about how Hitler came to power: his path to power was cleared by the same oligarchs who employ the brainwashers that program your television. Over a period of years, following the First World War, Germany was brutalised by the economic policy of this international elite. Hitler's Nazis were funded and promoted as a political option, and then steered into power in 1932-1933.

Once in power, the Nazis maintained their hold through the use of terror as part of mass brainwashing. In many ways it were proper to view the Nazi period as an experiment in methods of mass brainwashing and social control. At the root of this experiment was the desire to create a New World Order based on reversing a fundamental premise of western Christian civilization: that man is created as a higher and distinct species from animals, created in the image of the living God and by Divine grace, imparted the Divine Spark of reason."[5]

What makes man human is our power of reason. The only thing greater than life is the power of the human mind. This is how mankind is measured. What separates us from animals is our ability to discover universal physical principles. It allows us to innovate, which subsequently improves the lives of people. Development of mankind, the development of the power of the individual and the nation depends upon scientific developments, upon seeking and discovering the truth as our highest goal, thus perfecting our existence. Truth always lies in the higher order of processes. True sovereignty lies not in popular opinion, but in the creative powers of the individual human mind.

So, it's a moral problem. Problem of mankind's destiny. As Lyndon LaRouche often stated, "every generation must advance beyond that of the preceding generation. And that hope, that that will happen, should be what's on the mind of the person who is dying of old age: that their life has meant something, because it laid the foundation for a better life than they knew."[6]

There is a fundamental clash of ideals: Those who measure up to a Renaissance view of a man and those who see themselves, by birthright above other men, who see mankind as an animal, whose worst impulses must be repressed by the state. This is the view of the "Enlightenment movement," and in its extreme form, the fascist state. "For mass brainwashing to work it must attack the Renaissance view of man,"[7] for no person with a strong moral compass seeking Truth can be brainwashed.

Until we can bring mankind into a real Age of Reason, into the age of progress and human knowledge, history will be shaped in actuality, not by the wills of masses of humanity, but by the mere handfuls who, for purposes of good or evil, steer the fate of mankind generally as herds of cows are steered to and from the pasture – and, occasionally, also to the slaughter-house.

FREUDIAN MASS BRAINWASHING

As much as it may surprise many, Nazi Germany was an experiment in Freudian mass psychology. What it means is that both Freud and the Nazis shared the belief system of man as a sinful beast who is allowed to exist under strictly imposed laws.

Man is not made in the image of the living God, says Freud; man has made god in his image, for the purpose of easing the pain of his existence. Freud called past intellectuals "infantile" for their defense of religious doctrine: "We shall tell ourselves that it were very nice of there were a God who created the world and was a benevolent Providence...Yet you defend the religious illusion with all your might. If it becomes discredited – and indeed the threat to it is great enough – then your world collapses. There is nothing left for you but despair of everything, of civilization, of the future of mankind. From that bondage, I am, we are free. Since we are prepared to renounce a good part of our infantile wishes, we can bear it if a few of our expectations turn out to be illusions."[8]

Freudian psychology, as preached by either Freud or by neo-Freudians like Carl Jung, became the rage in the 1920s. It was promoted into the popular culture through the mass media of its day, newspaper and magazine articles. "It's morally insane system of 'id', 'ego' and 'superego' became part of the popular culture as did its belief that creativity stems from sexual drives."[9]

One of the key elements of Freud's work on mass psychology stems from *The Psychology of the Crowd* of the French psychologist Gustav LeBon who claimed that man regressed to a primitive mental state as part of a crowd. "People in a crowd lose their inhibitions and moral standards, and become highly emotional."[10] LeBon describes this as a return to man's primitive nature. "Man," says LeBon, "has returned to his animal roots." This is what mass mobilization of people is all about. The victory parade is just one example of this phenomenon. The notion that all good citizens should rally around the flag to attack, is in itself just another form of social dictate.

But there is another side effect to this development. "While he has become at once more primitive, more animal-like and infantile, mass man, the man in the crowd, also has a heightened sense of power, while his individual responsibility for action – a key factor in all moral judgement – diminishes."[11]

On the other hand, far from accepting that everything should be directed toward a common purpose, Freud and LeBon ignore the right of minds to follow their own curiosity wherever it may lead. After all, curiosity...is insubordination in its purest form, an attitude that in today's United States would draw you a prison term for anti-American activity. The heritage of western Christian civilization defends the freedom of the individual mind against the coercion of mass culture, mass propaganda or mass mobilization.

"It is just as impossible to do without control of the masses by a minority as it is to dispense with coercion in the work of civilization," Freud writes in his 1927 attack on religion, *The Future of an Ilusion*. "For the masses are lazy and unintelligent."[12] Such thoughts are not far separated from the writings of Hitler; they form the core of Nazi-thinking.

Long before Hitler published *Mein Kampf*, Freud, in his *Mass Psychology* wrote about a leadership principle around which the Nazi State was organized. "Any mass, be it a nation, or a randomly created group, must have a leader," he wrote, adding "someone who gives it its 'I ideal' or value. The leader becomes the individual member's common 'I ideal' and takes over all his critical faculties, just as the hypnotised individual surrenders his self-determination to the hypnotiser. It is the leader," says Freud, "who provides the common bond for a mass of people; their common attachment to the leader, enables each member to identify with each other, giving form and direction to the mass."

Was Hitler a Freudian? "It is known that Hitler read LeBon," writes Lonnie Wolfe in his piece on "The Making of a Fascist Society": "It cannot be established that he read Freud, especially *Mass Psychology*. But it is clear that those who put Hitler in power and those who steered his movement read Freud, as did most of the ruling elite of the day: it was they who were promoting the Freudian craze and its propaganda throughout the world."[13]

CARL JUNG AND HITLER

One key neo-Freudian who became an overt supporter of the Nazis was the Swiss psychoanalyst Carl Jung, whose friendship with Freud ended over the latter's refusal to see value in Gnostic mysticism. Freud, who was opposed to integrating mystical ideas into psychoanalysis, associated the word mysticism with séances, voices from other worlds, noises, apparitions, levitation, trances and prophecies.

"Jung saw in Hitler the apotheosis of Jung's own search for a kind of pagan communion with the Beyond, a search that began in 1915, with Jung's colossal nervous breakdown."[14]

In his 1997 essay on the subject of Hitler and Jung, Wolfe believes there is the strongest possible connection between Jung's psychoanalytic theories, which form one of the conceptual bases of "New Age" ideology today, and his fascination with Hitler. "For Jung was obsessed by the notion that the deepest reality, the greatest truth, lay buried in the unconscious, mystical, psychotic aspects of Man's mind, as opposed to the rational, outward, scientific, Judeo-Christian view of the world."[15] That was the basis of Jung's decades long search through himself, "attempting to find a pre-existing myth or mythic system which aptly illustrated his ideas about the human psychology of religion."[16] He began with Gnosticism, then went on to study astrology and then speculative alchemy as a symbolic system.

To Jung, there is a deep substratum of consciousness that lies beneath the layers of mechanical instincts and the measurable phenomena of clinical psychology. Jung called it the "collective unconscious." These images become visible under certain circumstances, such as in political rallies or religious rituals or on the movie screen or in advertising and propaganda, and we take them for granted without realizing the power they represent or the extent to which they may be manipulating our consciousness. And in turn, these images, this pattern or matrix underlying the observed universe, a kind of grid of connections linking events according to a system we can only barely perceive, are manipulated by the unseen hand of the brainwashers at Tavistock Institute and Frankfurt School.

Tavistock Institute for Human Relations is the psychological warfare arm of the British Royal family, located in a suburb of London. It is the world's most important institution for manipulation of population. According to the official history of the Tavistock Clinic: "In 1920, under its founder Dr. Hugh Crichton-Miller's leadership, the Clinic made a significant contribution to the understanding of the traumatic effects of 'shell shock'"[17] In the 1930s, the Tavistock Institute developed a symbiotic relationship with the Frankfurt Institute for Social Research. Their collaboration led them to analyze the culture of a population from a neo-Freudian standpoint. Nazism, just happened to have been one of its 'patients on a psychiatric couch.'

In a nutshell, Hitler was the prototype of Jungian man, who surrendered his reason to his unconscious, and who welcomed divine madness as Jung himself advised.

Jung had been impressed by Hitler's meteoric rise to power and recognised that the dictator "must have tapped some extraordinary energy in the Teutonic unconscious."[18] For example, in March 1934, Jung was writing of the "formidable phenomenon of National Socialism on which the whole world gazes with astonishment[19] Jung goes on to suggest that the Aryan unconscious has a "higher potential than the Jews." Hitler, Jung wrote, "had literally set all Germany on its feet." In a 1932 essay, Jung had celebrated the "leader personality" of the Fuehrer, as against the "ever-secondary, lazy masses, who cannot take the least move in the absence of a demagogue." In 1936, Jung perceived Hitler to be of the previously repressed Wotanic (Ancient Germanic God) elements: "The impressive thing about the German phenomenon is that one man, who is obviously possessed has infected a whole nation to such an extent that everything is set in motion and has started rolling on its course towards perdition."[20]

Was Jung unaware of the Nazi's true nature? Hardly. In 1938, five years after Hitler's rise to power, Jung classified Hitler as "visionary" and "a truly inspired shaman or medicine man" whose power was "magical rather than political," a "spiritual vessel," the "first man to tell every German what he has been thinking and feeling all along in his unconscious about German fate."[21] Jung added that "Hitler's power is not political, it is magic."

As France surrendered to Germany in June 1940 – the date, the summer solstice, did not pass unnoticed by Jung and other Nazi mystics – Jung cried ecstatically, "It is the dawning of the Age of Aquarius!" It is rather fine irony that it was Carl Jung, of all people, who coined that phrase made current so much later, with the rise of the "New Age."

Needless to say, the Nazis represented but a minority of the German population, even when they were in power. What about the so-called "good Germans" who went along with Hitler's terror? How was it done? The same way, it is being done to us today, and every day, through the mass media dissemination of "information."

And the most universal of the mass media was radio. In fact, once in power, the Nazis ordered the production and mass dissemination of cheap radio receivers for the entire population of the country. This is the real meaning of "mass audience." The concept behind it was the same as discussed by Freud in *Mass Psychology*, that is "individuals participating in the mass phenomenon are susceptible to suggestion, to losing his moral conscience – thus become overwhelmed by the mass experience." Brainwashers call it "institutional human aggressiveness," what people like Freud say proves that people, us, are animals driven towards destruction. According to Freud, these aggressive destructive drives are "part of man's animal nature." The purpose of society, according to Freud, is to "regulate and control through various forms of coercion the outbursts of this innate bestiality against which the human mind is ultimately powerless." Freud's principal point was that "masses of people can be organized around appeals to the emotions. The most powerful such appeals are to the unconscious, which has the power to dominate and throw aside reason."

"The mass has never thirsted for truth," Freud wrote, referring to crowds of raw minds swayed by elementary needs. "They demand illusions and cannot do without them. They constantly give what is unreal precedence over what is real; they are almost as strongly influenced by what is untrue as what is true. They have an evident tendency not to distinguish between the two."

Freud, in his *Mass Psychology* stated that an individual's moral inhibitions and outlook can be broken down as parts of a mass. Lonnie Wolfe succinctly tells the story of Freud's so-called powerful mass experience which can gratify emotions: "Freud further stated that under this condition, with man's reason dominated by emotionalism and unable and unwilling to look for Truth, the individual in a crowd loses his moral conscience, or what Freud calls his ego ideal." To Freud, this is not necessarily a bad thing, "since the moral conscience or the superego, causes man to unnaturally repress his basic animal instincts; This Freud claims produces neurosis. In a crowd organized around people's emotions, the individual will exhibit a tendency to 'let himself go,' to free himself of all moral and social inhibitions. Isolated, he may be a cultivated individual; in a crowd, he is a barbarian – that is creature acting by instinct." Therefore, the key to mass brainwashing is to create an organized, controlled environment in which "stress and tension can be applied to break down morally informed judgement, thereby making an individual more susceptible to suggestion."[22]

The point to be made here is that co-participation in a crowd causes a person to respond to situation from an emotionally-non-thinking set of reference points.

In the case of Nazi Germany, coming across the radio, into millions of homes, was the voice of one man, Adolf Hitler. The fact that all of Germany was hearing his voice at the same time gave an enhanced power to the message. The listener was literally part of a mass-experience, taking it all in from an emotionally-non-thinking set of reference points. Hitler's speeches were some of the first mass-media events in history – as carefully staged as any event in modern history.

Both Tavistock and the Frankfurt School paid close attention to Nazi propaganda techniques, which they willingly incorporated into their research. The aim of this project, as stated in Adorno's *Introduction to the Sociology of Music*, was to "program a mass culture as a form of extensive social control that would steadily degrade its consumers." The application of their research into human behaviour was set to launch a decade later in a major irreversible Cultural Revolution in America.

"The brainwashers concluded that mass media events had caused people to suspend their belief in reality, that they had in fact been willing to accept uncritically things being said, which if they had heard in another context they would most probably had rejected." Now, think back to today. How insane are some of the things we have been told by our leaders? Weapons of mass destruction in Iraq,

Iranian mullahs allegedly threatening the security of the United States, Libyan leader Mu'ammar al-Qaddafi supplying his troops with viagra in order to rape women participating in the rebellion, Usama bin Laden's death. Explains Wolfe: "During the Second World War, Bruno Bettleheim, a neo-Freudian, published a psychological analysis of the Nazi period at the behest of the network of brainwashers associated with the Tavistock Institute. Bettleheim describes how under extreme doubt and terror, the individual will regress to an increasingly more infantile state. In that condition, the inmates of the nazi concentration camps started to mirror the personalities and mannerisms of their oppressors, the SS guards. In a widely circulated version of his work, *The Informed Heart*, he indicated that life outside the concentration camps mirrored the psychological disintegration taking place inside: all German citizens were becoming more infantile, less able to act as reasonable adults."[23]

In *The Informed Heart*, Bettleheim writes: "the 'good German' had to be unseen and also dumb...It is one thing to behave like a child because one is a child...It is quite another thing to be an adult and have to force oneself to assume childish behaviour. It was not just coercion by others into helpless dependency. It was also the clean splitting of the personality. Man's anxiety, his wish to protect life, forced him to relinquish what was ultimately his best chance of survival: his ability to react and make appropriate decisions. But giving this up, he was no longer a man, but a child. Knowing that for survival, he should decide and act, and trying to survive by not reacting – these in their combination overpowered the individual to such a degree that he was eventually short of all self respect and all feelings of independence."[24]

In the end, the Nazi experiment failed and the Nazis themselves, a group of gnostic psychotics, had to be destroyed by the same forces who put them in power in the first place. In the meantime, the Third Reich caused a massive shift of consciousness in the planet, a paradigm shift, if you like. Although occultism and politics have long been bedfellows, since the days of Joseph interpreting dreams for Pharaoh, or King Saul consulting the witch of Endor, or even earlier in the astral temples at Nineveh and Babylon, the world came very close to enduring a massive and industrialised 'cult-ocracy' in the form of the Third Reich. It is worth repeating here that the Nazi Party was not a political party, as we commonly understand it, but a cult.

"Nazi appropriation of the occult was a weird farrago of astrology, Freemasonry, racism rooted in occultism and popular European folklore (the Cathars, the Holy Grail, the Knights Templar, the Arthurian legends). Sinister Forces at work against humanity. In *Morning of the Magicians*, one of the most explosive books of the 1960s, the authors identified these forces in paranormal terms, but always linking them back to the fascist fantasies of Hitler, Himmler, Rosenberg, Darre and Hess: blatant occult fantasies that became government policy and which led to the Holocaust and World War II."[25] That millions of innocent people died because of this twisted dream is the greatest tragedy of the XX century. But the authors insisted that there was something there, below the surface of the Third Reich: a sinister force that had been successfully, for a time, evoked by the Nazi magicians of the SS. The phenomenon of 'coincidence' or what Carl Jung called "synchronicity" was the most obvious evidence of the operation of this force and one which the authors suggested could be the basis of a whole "new conception of history."[26]

And when it was all over, those who imposed this horror on the world attempted through mass media to blame their victims for the crimes committed. The Germans, whom the oligarchy through their Nazi tools, had tortured in mass brainwashing were told that they were collectively guilty for all that happened. As a result, the entire nation was put in the dock, and tried as criminals, rapists and mass murderers. And while we were told that it must never happen again, the brainwashers at such places as Tavistock Institute were already secretly working behind the scenes on a new and far more powerful brainwashing tool – television, to help them organize their fascist 'superstate' without the now socially unacceptable Nazi superstructure.

Now, how many people actually understand this? How many people realize that the majority's perception of reality, especially in political arenas, is not their own. It is shrewdly manipulated and imposed upon them by the Men behind the Curtain. Many certainly do not and brainwashing has a lot to do with it.

CONTROLLING THE ENVIRONMENT

What became obvious to the brainwashers is that a new system, based on appeal to emotionalism, was needed. And in order to break through the population's moral compass, society needed to be reduced to infantilism. Writing in 1972, Tavistock's leading media expert, Dr. Fred Emery, reported on television's impact on Americans: "We are suggesting that television evokes a basic assumption of dependency. It must evoke (this) because it is essentially an emotional and irrational activity...television is the non-stop leader who provides nourishment and protection."

In his informative report on television's impact on cognitive powers of an individual, investigative journalist, Lonnie Wolfe says that both Emery and Eric Trist, who until his death in 1993 headed Tavistock's operations in the United States, noted that "all television had a dissociative effect on mental capabilities, making people less able to think rationally. Viewers, as they become habituated to watching six hours or more of television daily, surrender their powers of reason to the images and sound coming from the tube."[27] Tavistock recognised that habituated television watching destroys the ability of a person for critical cognitive activity. In other words, it makes you stupid.

For to bring society down to the level of a beast is especially important from the point of view of Tavistock, if they are to control the planet Earth. "Since the only source of increase of mankind's power, as a species, within and upon the universe, is that manifold of validated discoveries of physical principle, it follows, that the only form of human action that distinguished man from beast, is that form of action, which is identified as cognition, by means of which the act of discovery of accumulated validatable universal physical principles is generated. It is the accumulation of such knowledge for practice, in this way, from generation to generation, which defines the provable evidence of the absolute difference between man and beast."[28]

However, dissociative effect is but one of the aspects of overcoming resistance and changing the established paradigm of society, according to brainwashers at Tavistock Institute. Listen to Theodore Adorno of Frankfurt School: "It seems obvious, that the modification of the potentially fascist structure cannot be achieved by psychological means alone. The task is comparable to that of eliminating neurosis, or delinquency, or nationalism from the world. These are products of the total organization of society and are to be changed only as that society is changed. It is not for the psychologist to say how such changes are to be brought about. The problem is one, which requires the efforts of all social scientists. All that we would insist upon is that in the councils or round tables where the problem is considered and action planned the psychologist should have a voice. We believe that the scientific understanding of society must include an understanding of what it does to people, and that it is possible to have social reforms, even broad and sweeping ones, which though desirable in their own right would not necessarily change the structure of the prejudiced personality. For the fascist potential to change, or even to be held in check, there must be an increase in people's capacity to see themselves and to be themselves. This cannot be achieved by the manipulation of people, however well grounded in modern psychology the devices of manipulation might be.... It is here that psychology may play its most important role. Techniques for overcoming resistance, developed mainly in the field of individual psychotherapy, can be improved and adapted for use with groups and even for use on a mass scale."[29]

SOCIAL TURBULENCE

As a point of interest, at Tavistock, Eric Trist and Frederick Emery developed a theory of "social turbulence," a so-called "softening up effect of future shocks" – wherein a population could be softened up through mass phenomena such as energy shortages, economic and financial collapse, or terrorist attack. "If the 'shocks' were to come close enough to each other and if they were delivered with increasing intensity, then it was possible to drive the entire society into a state of mass psychosis," claimed Trist and Emery. They also stated that "individuals would become disassociated, as they tried to flee from the terror of the shocking, emerging reality; people would withdraw into a state of denial, retreating into popular entertainments and diversions, while being prone to outbursts of rage."

In fact, we are talking about two sides of the same coin here. On one side, guiding the covert, subtle manipulation and control of thought and human consciousness through the power of television; while on the other side, directly and overtly shifting the paradigm, changing the basic concepts, widening the parameters, and changing the rules by which society defines itself.

One of the key individuals involved in psychological warfare against the population through manufactured social turbulence is Kurt Lewin, a pioneer in group dynamics who was part of the early Frankfurt School and fled Germany when Hitler took power. This passage from his book *Time Perspective and Morale*, shows his understanding of psychological warfare: "One of the main techniques for breaking morale through a 'strategy of terror' consists in exactly this tactic – keep the person hazy as to where he stands and just what he may expect. If in addition frequent vacillations between severe disciplinary measures and promises of good treatment together with spreading of contradictory news, make the 'cognitive structure' of this situation utterly unclear, then the individual may cease to even know when a particular plan would lead toward or away from his goal. Under these conditions even those who have definite goals and are ready to take risks, will be paralyzed by severe inner conflicts in regard to what to do."[30]

Over the past 50 years, research in the fields of psychology, sociology and psychiatry has shown that the human mind is capable in dealing with a limited number of changes around it. According to Science Policy Research Unit [SPRU] at Tavistock's Sussex University facility, 'future shocks' is defined "as physical and psychological distress arising from the excess load on the decision-making mechanism of the human mind." In other words, "a series of events, which come so fast that the human brain cannot absorb the information." One scenario is called superficiality. After continuous shocks, according to Emery and Trist, "the large targeted population group discovers that it does not want to make choices any more, reducing the value of his intentions... This strategy can only be pursued by denying the deeper roots of humanity that bind...people together on a personal level by denying their individual psyche."

Apathy takes over, often preceded by mindless violence such as is characteristic of the Los Angeles street gangs in the 1960s and the 1980s, what Emery and Trist call organized social response to dissociation, as described in the pages of Anthony Burgess' novel *A Clockwork Orange*, a society dominated by infantile animal-like rage. "Such a group becomes easy to control and will docilely follow orders without rebelling, which is the object of the exercise,"[31] add Trist and Emery. What's more, "the dissociated adults cannot exert moral authority over their children, because they are too involved with their own infantile fantasies, brought to them through their television set."[32] And if you doubt what I am saying, look at the older generation today as they have accepted the moral decadence of the no-future generation of its children, rather than seek conflict, and in the process, the adults have come to accept a lower moral standard.

Just as in Huxley's drug-controlled *Brave New World*, there are no moral or emotional choices to make here, the 'flower children' and the drug-soaked rebellion of the Vietnam era is a perfect example of how this scenario functions.

These 'frequent vacillations' pass through several scenarios: Stable, at which point, people more or less are able to adapt to what is happening to them, or it is turbulent, "at which point people either take actions to relieve the tension, or they adapt to accept tension-filled environment."[33] If the turbulence does not cease, or is intensified, then "at a certain point people cease being able to adapt in a positive way."[34] According to Trist and Emery, "people become maladaptive – they choose a response to tension that degrades their lives. They start to repress reality, denying its existence, and constructing increasingly more infantile fantasies that enable them to cope. Under the conditions of increasing social turbulence, people change their values, yielding to new degraded values, values that are less human and more animal-like."[35]

The second scenario is segmentation of society into smaller parts. In this scenario, "it is every group, ethnic, racial, and sexual against the other."[36] Nations break apart into regional groups, those smaller areas in turn fissure into even smaller areas, along ethnic lines. Trist and Emery refer to it as "enhancement of in-group and out-group prejudices as people seek to simplify their choices. The natural lines of social divisions emerge to become barricades."

Society's response to such a psychological and political disintegration is the Orwellian fascist state, modelled on his book *1984*. In the book, "Big Brother" regulates the lives and conflicts of people within a society; a never ending conflict "is waged by each ruling group against its own subjects, and the object of the war is not to make or prevent conquest of territory, but to keep the structure of society in tact."[37]

"The third scenario is the most intense, involving a withdrawal and retreat into private world and a withdrawal from social bonds that might entail being drawn into the affairs of others."[38] How is this level of dissociation different from today's 15-24 age group? How far removed are we from this moral and social outlook? Almost there, aren't we? Trist and Emery are convinced that men will be willing to accept "the perverse inhumanity of man that characterised Nazism." Not necessarily the structure of the Nazi state, but the moral outlook of Nazi society.

To survive in such a state, people will need to create a new religion. Wolfe states that "the old religious forms, especially western Christianity, demand that man be responsible for his fellow humans. The new religious forms, will be a form of mystical anarchism, a religious experience much likened to satanic practice of the Nazis or the views of Carl Jung."[39] This is the "New Age," the "Age of Aquarius" preached by Tavistock and Frankfurt School with its eastern, mystical religious cults that had young, brainwashed converts flocking to embrace this degeneracy. Again, it is television that provides the "social glue" that binds the minds of the population to their new religious forms.

NEWSSPEAK

Most Americans and Europeans alike believe that there is such a thing as a free press, one of the key areas in brainwashing the population. Most Americans and Europeans get most of their news from state-controlled television, under the misconception that reporters are meant to serve the public. In fact, reporters do not serve the public. Reporters are paid employees and serve the media owners, whose company's shares are traded on Wall Street.

Of the six to eight hours per day Americans spend watching television,[40] for example, a couple of hours is spent watching current affairs programming. Control over the media has been a long-term objective of the globalist elite. In February 1917, Congressman Oscar Callaway placed a statement in the Congressional Record describing the origins of what he called the "newspaper combination." According to that account, the J.P. Morgan Banking interests and their allies "got together 12 men high up in the newspaper world and employed them to select the most influential newspapers in the United States and the sufficient number of them to control generally the policy of the daily press in the United States." Today, the *New Media Monopoly* puts the number at five corporations controlling America's Fourth Estate.

As I explain in my book on the Bilderberg Group, in 1993 August/September edition, the prestigious Dutch magazine *Exposure* outlined disturbing details about how the Tavistock Institute for Behavioural Analysis, planned to control the boards of the three major and most prestigious television networks in the United States: NBC, CBS and ABC. All three television networks came as spin-offs from the Radio Corporation of America (RCA). These organizations and institutions that theoretically are in "competition" with each other – this is part of the "independence" that ensures Americans enjoy unbiased news – are in fact closely interfaced and interlocked with countless companies and banks, making it an almost impossible task to untangle them.

Take NBC, for example. NBC has been affiliated with RCA, a media conglomerate, and is now a subsidiary of Morgan-controlled General Electric. On RCA's board sits Thornton Bradshaw, president of Atlantic Richfield Oil, and member of the World Wildlife Fund, the Club of Rome, the Aspen Institute for Humanistic Studies, and the Council on Foreign Relations. Bradshaw is also chairman of NBC.

Council on Foreign Relations is part of the Royal Institute for International Affairs. RIIA, is a front organization for The Round Table Group, "stemmed from a secret society created by British magnate Cecil Rhodes to unite the world – beginning with the English-speaking dominions – under 'enlightened' elitists like himself." (Quigley, *Tragedy and Hope*) This front organization, the Royal Institute for International Affairs, had as its nucleus in each area the existing submerged Round Table Group. In New York it is known as the Council on Foreign Relations.

RCA's most legendary role, however, was the service it provided to British Intelligence during World War II. Of particular note: RCA's President David Sarnoff moved to London at the same time Sir William Stephenson (of Intrepid fame) moved into the RCA building in New York. Today, RCA's directorate is made up of British-American establishment figures that belong to other organizations such as the CFR, NATO, the Club of Rome, the Trilateral Commission, Bilderberg, Round Table, etc.

Among the NBC directors named in the *Exposure* article were John Brademas (CFR, TC, Bilderberg), a director of the Rockefeller

Foundation; Peter G. Peterson (CFR), a former head of Kuhn, Loeb & Co. (Rothschild), and a former U.S. Secretary of Commerce; Robert Cizik, chairman of RCA and of First City Bancorp, which was identified in Congressional testimony as a Rothschild bank. Clearly the NBC board is considerably influenced by the Rockefeller-Rothschild-Morgan troika, leading exponents of the One World Company Ltd.

ABC is owned by Disney Corporation. It has 153 TV stations. Chase Manhattan Bank controls 6.7% of ABC's stock – enough to give it a controlling interest. Chase, through its trust department, controls 14% of CBS and 4.5% of RCA. Instead of three competing networks called NBC, CBS and ABC, what we really have is the Rockefeller Broadcasting Company, the Rockefeller Broadcasting System, and the Rockefeller Broadcasting Consortium.

On the ABC board of directors is Ray Adam, director of J.P. Morgan, Metropolitan Life (Morgan), and Morgan Guarantee Trust; Frank Cary, chairman of IBM, and director of J.P. Morgan and the Morgan Guarantee Trust; John T. Connor (CFR) of the Kuhn Loeb (Rothschild) law firm; George Jenkins, chairman of Metropolitan Life (Morgan) and Citibank (Rothschild connection); Martin J. Schwab, director of Manufacturers Hanover (Rothschild).

Isn´t it strange how the same Rockefeller-Rothschild-Morgan characters on the board of the ABC network, which we are told, is independent of NBC, appear to represent the competition? ABC was taken over by Cities Communication, whose most prominent director is Robert Roosa (CFR, Bilderberg), senior partner of Brown Brothers Harriman, which has close ties with the Bank of England. Roosa and David Rockefeller are credited with selecting Paul Volcker to chair the Federal Reserve Board. Selection of Volcker formed part of Tavistock's plan for a permanent condition of social turbulence, such as the CFR's Project 1980s and its "controlled disintegration" of the U.S. economy: a pre-planned collapse, carried out by Volcker and his high interest rate policy of the 1979-82 period.

CBS is owned by Viacom, which has over 200 TV and 255 radio affiliates nationwide. This huge media conglomerate owns, among other companies, MTV, Showtime, VH1, TNN, CMT, 39 broadcast television stations, 184 radio stations and Paramount Pictures. As an American intelligence officer, CBS founder William Paley was trained in mass brainwashing techniques during World War II at the Tavistock Institute in England.

The financial expansion of CBS was supervised for a long time by Brown Brothers Harriman and its senior partner, Prescott Bush (father and grandfather to Presidents), who was a CBS director. The Board includes Chairman Paley, for whom Prescott Bush personally organized the money to buy the company; Harold Brown (CFR), executive director of the Trilateral Commission, and former Secretary of the Air Force and of Defense of the U.S.; Roswell Gilpatric (CFR, Bilderberg) from the Kuhn, Loeb (Rothschild) law firm; Henry B. Schnacht, director of the Chase Manhattan Bank (Rockefeller/ Rothschild), CFR, Brookings Institute; Franklin A. Thomas (CFR), head of the Rockefeller-controlled Ford Foundation; Newton D. Minor (CFR), director of the RAND Corporation and, among many others, the Ditchley Foundation, which is closely linked with the Tavistock Institute in London and the Bilderberg Group. The former president of CBS was Dr. Frank Stanton (CFR), who is also trustee of the Rockefeller Foundation and Carnegie Institution. So are the Rothschild and Rockefeller families, who are the leading groups in the tightly controlled field of communications working closely with Tavistock Institute in London.

Fox News, part of News Corp, with their daily Nuremberg rallies for couch potatoes, is owned by Rupert Murdoch, who owns a significant part of the world's media, amongst them his flagship publication, the Wall Street Journal, MySpace and 20th Century Fox movie studio. Murdoch's Media Empire has been the main propaganda outlet for the perpetual war of the neo-cons and their Nazi minions.

Then, there is the Corporation for Public Broadcasting. PBS, supposedly a public institution. According to its web site, "PBS, is a non-profit media enterprise owned and operated by the nation's ... public television stations. A trusted community resource, PBS uses the power of non-commercial television, the Internet and other media to enrich the lives of all Americans through quality programs and education services that inform, inspire and delight. Available to 99 percent of American homes with television and to an increasing number of digital multimedia households, PBS serves nearly 90 million people each week."[41]

The cornerstone of PBS's programming was, until a few years ago, the evening television news program, *The NewsHour with Jim Lehrer*. Jim Lehrer is a member of the Council on Foreign Relations. For most of PBS's history, funding was provided by AT&T (A CFR company); Archer Daniels Midland, whose Chairman Dwayne Andreas was a member of the Trilateral Commission; PepsiCo (a CFR company), whose Chief Executive Officer, Indra Krishnamurthy Nooyi, is a Bilderberg and Trilateral Commission Executive Committee member; and smith Barney (a CFR company), one of the world's leading financial institutions. Furthermore, Smith Barney is interlocked with Citigroup Inc., a global financial services company that is a member of the Bilderberg Group, CFR and the Trilateral Commission.

Journalists who participate in *The NewsHour* are some of the best-known political pundits in the United States, such as Paul Gigot, David Gergen, William Kristol and William Safire. All of them are members of Bilderberg Group, CFR or the Trilateral Commission.

Would it be reasonable for us to suspect that PBS might not be quite as impartial in certain delicate matters of public interest, such as the U.S. Constitutional crisis envisioned by Brzezinski, the future of nation-state republics, and of national sovereignty?

Still not convinced? "American media elites practice a brutal, albeit well-concealed, form of 'wartime' news censorship, but the mechanisms of this control are now openly acknowledged. John Chancellor, the longtime NBC-TV news anchorman, in his recent autobiographical account of life in the news room, *The New News Business*,[42] admitted that, through formal structures such as the Associated Press, informal 'clubs' such as the New York Council on Foreign Relations, decisions are made, on a daily or weekly basis, about what the American people will be told, and what stories will never see the light of day."[43]

David Rockefeller specifically identified the *New York Times* and the *Washington Post* as key media organs of the power elite.

As Michel Chossudovsky explains on the pages of globalresearch. com, "historically, the *New York Times*, with utterly unwarranted self-assurance, designates itself the arbiter of 'All the News that's Fit to Print,' has served the interests of the Rockefeller family in the context of a longstanding relationship. The current *New*

York Times chairman Arthur Sulzberger Jr. is a member of the Council on Foreign Relations, son of Arthur Ochs Sulzberger and grandson of Arthur Hays Sulzberger who served as a Trustee for the Rockefeller Foundation. Ethan Bronner, deputy foreign editor of the *New York Times* as well as Thomas Friedman among others are also members of the Council on Foreign Relations (CFR)."[44]

America's corporate media are an integral part of the economic establishment, with links to Wall Street, the Washington think-tanks, Club Bilderberg and the Council on Foreign Relations (CFR) and through them to the world's premier brainwashing center, Tavistock Institute. CFR is "the premier U.S. foreign policy think-tank in the United States, and is one of the central institutions for socializing American elites from all major sectors of society (media, banking, academia, military, intelligence, diplomacy, corporations, NGOs, civil society, etc.), where they work together to construct a consensus on major issues related to American imperial interests around the world. As such, the CFR often sets the strategy for American policy, and wields enormous influence within policy circles, where key players almost always come from the rank and file of the CFR itself."[45]

CFR's former President, Sir Winston Lord, made an interesting observation when he said that "The Trilateral Commission doesn't secretly run the world. The Council on Foreign Relations does that."[46]

In his October 30, 1993 "Ruling Class Journalists" essay, *Washington Post* ombudsman Richard Harwood candidly discussed how the CFR dominates the news media. Harwood described it as the "the closest thing we have to a ruling Establishment in the United States.... Its members are the people who, for more than half a century, have managed our international affairs and our military-industrial complex." After listing the executive branch positions then occupied by CFR members, Harwood continued: "What is distinctly modern about the council these days is the considerable involvement of journalists and other media figures, who account for more than 10 percent of the membership.

"The editorial page editor, deputy editorial page editor, executive editor, managing editor, foreign editor, national affairs editor, business and financial editor and various writers as well as (the now deceased) Katherine Graham, the paper's principal owner, represent the *Washington Post* in the council's membership," observed

Harwood. These media heavyweights "do not merely analyse and interpret foreign policy for the United States; they help make it," he concluded. Rather than offering an independent perspective on our rulers' actions, the Establishment media act as the ruling elite's voice – conditioning the public to accept, and even embrace "insider" designs that otherwise might not be politically attainable.

PROPAGANDA CAMPAIGN

Even in the age of internet, blogaspheres and other new generation contraptions of mass dissemination of information, the *Times* and the *Post*, both Bilderberg and CFR companies, set the tone for most news coverage, defining issues and setting the limits of "respectable" opinion.

Consider, for example, Sudan or Libya to illustrate how the mainstream ruling class journalists "help make" foreign policy through a blatant propaganda campaign.

"Saturation reporting from a crisis region; emergency calls for help broadcast on the electronic media (such as the one recently on the CFR run PBS); televised pictures of refugees (such as the one recently on CFR run Fox and CNN); lurid stories of 'mass rapes,' which are surely designed to titillate as much to provoke outrage; reproachful evocations of the Rwandan genocide; demands that something must be done ('How can we stand idly by?', etc.); editorials (in the CFR run *New York Times, Washington Post, Newsweek, Time* magazine and in CFR's own *Foreign Affairs* magazine with long winded opinion pieces by Bilderberg-CFR member Zbiegniew Brzezinski), calling for a return to the days of Rudyard Kipling's benevolent imperialism; and, finally, the announcement that plans are indeed being drawn up for an intervention."[47]

Now, let's turn on the television and see for ourselves how the news is packaged for our consumption. All 'news items' are told in short bits, most running no longer than 30 seconds; a major news item may run up to a minute or a minute and a half. Voice over pictures. Short interviews, usually only a few sentences, a few snippets of disjointed thought. The average half-hour segment may report up to 40 items, all presented in a seamless style, followed by sports, weather, entertainment and idle banter between the newsreaders.

But, are we witnessing real events from the real world? Is that how complex world issues are seen in situ? Libya shoehorned into a 30-second sound bite? Afghanistan's 2,000 years of history into a 45-second interview dissected into three shorter interview bits, each consisting of three sentences of six words? Or is it a badly distorted picture that the newsreader tells you is reality?

Given the way the news is presented, does your mind ever engage in deliberative thought about any single item? No, it doesn´t. Rather, you watch, merely take in the information, in the format it is presented to you. Would it surprise you then, if I told you that within two hours, most viewers could only remember vague summary about what they saw? Middle East is on fire. More civilians were killed somewhere. Floods wiped out a large chunk of land in...yes, where? European economy is weak and another country has asked to be rescued. There was an earthquake in Japan and miraculously someone survived a nine-day ordeal buried under rubble. Wow! Cool! Get that person on the Oprah Winfrey Show! Hi mom, I am on TV!

In his article on the "Cartelization of the News Industry," Jeffrey Steinberg explains that "even the live coverage of car crashes, gang murders, rapes, natural disasters, wars, terrorist acts, is served up on the basis of careful studies conducted at the neurological divisions of the leading medical schools. In recent decades, psychological warfare experts have unveiled a new pseudo-science called 'victimology', developed by the London Tavistock Institute, which is premised upon the theory that individuals can be put through trauma by being exposed to shockingly graphic visual accounts of violence."[48]

This type of brainwashing, according to Lonnie Wolfe, is called "selective retention. Television causes people to suspend their critical judgement capabilities because the combination of sound and images places the individual in a dream-like state, which limits cognitive powers."[49]

If you were to summarize the news items, what would you say? That we are living in a violent and degraded society. In other words, a degraded view of man as animal, killing, murdering, raping innocent people the world over. Secondary image: economic collapse, fear and hopelessness. You are left with this vivid image of society planted in your head. It is painted thus because no cognitive thought process was used to arrive at the real story behind the events.

Hal Becker from the Futures Group contends that through the control of television news programming, he can create popular opinion by manipulating the way you think and act. "Americans think they are governed by some bureaucrats in Washington who make laws and hand out money. How wrong they are. Americans are ruled by their prejudices and their prejudices are organized by public opinion... We think that we make up our minds about everything. We are so conceited. Public opinion makes up our minds. It works our herd instinct, like we are frightened animals."

But there is a very big difference between a beast and a man. This difference lies in our search for eternal Truth and for the meaning of Life. Truth always lies in the higher order of processes and in the creative powers of the individual human mind. So, it's a moral problem. The problem of mankind's destiny that no animal can ever solve. Every generation must advance beyond that of the preceding generation. And that hope, that that will happen, should be what's on the mind of the person who is dying of old age: that their life has meant something, because it laid the foundation for a better life than they knew.

On the other hand, people like Freud, Bertrand Russell, Eric Trist, Emery, Adorno who believe there is no significant difference between man and animal, must deny the existence and relevance of eternal Truth and our search for the meaning of Life, to render all men morally insane.

"Television, with its overwhelming presence in your life, both creates popular opinion and simultaneously validates it."[50] How is it done, exactly? That's because your identity, thanks to television, is shaped by what others think about them – "a constant and unending desire to act as you perceive others would want you to act."[51]

For example, "America is saying 'NO' to drugs," according to the latest polls by Gallup, says a smiling face of a young journalist on CNN. Is that right? Are we supposed to believe this nonsense just because a poll says so? Or because CNN says so? As Becker says, "the world is in that box. And it's there every night. Well, it's really there a lot more than that: 6-8 hours per day."

In a sense, the 'news' media generally tell one story: The saga of Government as Savior. On nearly every conceivable issue, domestic or foreign, news stories are designed to encourage readers and viewers to look to government intervention as a solution. In other words, the establishment media are guilty of conscious and willing accomplices in the power elite's drive for global control. Because, in order to control the world, the power elite must conquer the public mind. And as Walter Lippmann reminds us, "News and the truth are not the same thing..."

WIKILEAKS AND THE ROLE OF THE CORPORATE MEDIA

Let's have a look at one very recent example of media manipulation. Wikileaks. Wikileaks was started up in December of 2006. Wouldn't you say it was rather strange how an alleged upstart 'leak' site, was given star-studded attention from corporate media such as the *Washington Post* and *Time* magazine? What's more, *Time* magazine immediate did a lot more than give credibility to Wikileaks by giving the audience tips on how best to decipher the site in such a way that seemed eerily familiar to George W. Bush's rhetoric immediately after 9/11.

This is what *Time* magaizine had to say about it in January 2007. "By March, more than one million leaked documents from governments and corporations in Asia, the Middle East, sub-Saharan Africa and the former Soviet Bloc will be available online in a bold new collective experiment in whistle-blowing. That is, of course, as long as you don't accept any of the conspiracy theories brewing that Wikileaks.org could be a front for the CIA or some other intelligence agency."[52]

What makes this whole thing so incomprehensible is that the *Time* article was written before Wikileaks' first big "leak," in March of 2007. Why would *Time* even mention them if Wikileaks were two months away from their coming-out party? Why would they expose themselves in such a risky way, unless of course they know something we don´t. Are we to believe that *Time* magazine, an establishement media, would independently praise an alleged anti-establishment leak site which, in the near future will expose all kinds of "conspiracy theories" while at the same time telling its audience that these conspiracy theories circulating on the web are all crazy? Just so long as you believe the "correct" conspiracy theories, you'll be all right I guess.

Now, read the following from *Time* magazine. "Instead of a couple of academic specialists, Wikileaks will provide a forum for the entire global community to examine any document relentlessly for credibility, plausibility, veracity and falsifiability," its organizers write on the site's FAQ page.[53] "They will be able to interpret documents and explain their relevance to the public. If a document is leaked from the Chinese government, the entire Chinese dissident community can freely scrutinize and discuss it… "[54] Does this sound like a public relations propaganda piece written by the boys at Madison Avenue?

Sounds like a Wikileaks "How To" guide. Please understand you are being told what to think and how to interpret the information. Of course, in order to interpret it "correctly" you must have the right mindset. But, in order to get the unsuspecting public into the right mindset, Wikileaks needed the official blessing of the leading mainstream corporate media publications. Without the public's support, this little project looked doomed to fail before it even got off the ground. Is that why you had Ruppert Murdoch's public relations guy on the board of directors of Wikileaks? Is that the reason both *Time* magazine and the *Washington Post* had to come out with supportive articles about Wikileaks before anyone knew anything about them?

Please understand, mainstream news limits your understanding of the world by what it chooses not to report and ignore.

POLLING BY NUMBERS

In the early 20th Century, the words "propaganda" and "war" became synonymous with one another. This was no accident. It was all part of the effort to undermine the Judeo-Christian legacy through an 'abolition of culture.' Two men showed how to link the concepts of propaganda and war into one technique. One was Walter Lippmann, one of the most influential of political commentators of his time. It was Lippmann who in 1922 gave us the concept of "stereotype," which was basically a continuation of the Jungian concept of the archtype by other means. Stereotypes can be created, and manipulated, by the gurus of mass communication and psychological warfare. The idea is not to make the target think too clearly or too profoundly about the images received but instead to react, in a Pavlovian manner, to the stimulus it provides.

The other was Edward Bernays, Freud's nephew and one of the founders of public opinion manipulation techniques. In Bernays' own words: "It was, of course, the astounding success of propaganda during the [First World] war that opened the eyes of the intelligent few in all departments of life to the possibilities of regimenting the public mind….We are governed, our minds are molded, our tastes formed, our ideas suggested, largely by men we have never heard of. Whatever attitude one chooses to take toward this condition, it remains a fact that in almost every act of our daily lives, whether in the sphere of politics or business, our social conduct or our ethical thinking, we are dominated by a relatively small number of persons, a trifling fraction of our hundred and twenty million [U.S. citizens at the time], who understand the mental processes and social patterns of the masses. It is they who pull the wires, which control the public mind, and who harness old social forces and contrive new ways to bind and guide the world."[55]

During WWI, as a young lad, Bernays served on the U. S. Committee on Public Information (CPI), an ingeniously named American propaganda apparatus mobilized by the U.S. government in 1917 to sell the wonders of 'righteous' war to the unaware population as one that would 'Make the World Safe for Democracy.' Even though millions were killed on the battlefields of Europe, there was hardly any resistance to the carnage. By 1917, shortly before the U.S. entered the war, 94% of the British working class, an overwhelming majority of those who perished on the battlefields of Europe in the name of their King, had no idea what they were fighting for, other than the distorted image generated by the corporate media of the time that the Germans wanted to kill their King and destroy their country. Every war since then, including the current "war on terror," has used identical mass communications propaganda techniques.

"The manipulators then played upon this to undermine and distract the grasp of reality governing any given situation and, the more complex the problems of a modern industrial society became, the easier it became to bring greater and greater distractions to bear so that what we ended up with was that the absolutely inconsequential opinions of masses of people, created by skilled manipulators, assumed the position of scientific fact."[56]

Today, public opinion polls are completely intertwined with current affairs television shows that offer the public ready-made

McNews items in easy to swallow bite-size pieces. A 'scientific survey' of what people are said to think about an issue can be produced in less than twenty-four hours. Most of the polls announced by the world's leading media groups, such as CBS-NBC-ABC-CNN-Fox, the *New York Times*, the *Washington Post*, *Time, Newsweek, Financial Times, Wall Street Journal* in fact are co-ordinated at the National Opinion Research Center where, as much as it will amaze most people, a psychological profile was developed for every nation on Earth.

The idea of 'public opinion' is not new, of course. Plato warned against it in his *Republic* over two thousand years ago as did Alexis de Tocqueville on its influence over America in the early 19th century. But, until the 1930's, no one thought of using it to sway the masses towards a particular vision for decision-making.

As Michael Minnicino explains at length, "The belief that public opinion can be a determinant of truth is philosophically insane. It precludes the idea of the rational individual mind. Every individual mind contains the divine spark of reason, and is thus capable of scientific discovery, and understanding the discoveries of others. The individual mind is one of the few things that cannot, therefore, be 'averaged.' Consider: at the moment of creative discovery, it is possible, if not probable, that the scientist making the discovery is the *only* person to hold that opinion about nature, whereas everyone else has a different opinion, or no opinion. One can only imagine what a "scientifically-conducted survey" on Kepler's model of the solar system would have been, shortly after he published the *Harmony of the World:* 2% for, 48% against, 50% no opinion.

"Despite its unprovable central thesis of "psychoanalytic types," the interpretive survey methodology of the Frankfurt School became dominant in the social sciences and essentially remains so today. In fact, the adoption of these new, supposedly scientific techniques in the 1930's brought about an explosion in public-opinion survey use, much of it funded by Madison Avenue."[57] That's because opinions can be easily counted. You can ask a group of people how they 'feel' about something or whether they think that such and such a statement is true. You add the 'yes' and subtract the 'no' with the larger number becoming the consensus, which apparently represent the opinion of the 'majority'. Thus, "America is saying NO to drugs" is a manufactured consent by the pollsters on behalf of the financial elite who are, in fact, the world's biggest drug pushers as I have repeatedly shown in all my books. But, no matter how many people agree or disagree with something, it doesn't make it true. After studying American polling numbers, Emery and Trist have come to a conclusion that "people behave according to consensus." The masses of great unwashed, are so directed and influenced by the polls that "they wouldn´t dare break this social contract: they must do what others perceive that they should do; to do otherwise would cause psychological pain." They conclude that polls, "give meaning to their lives."[58]

After World War II, Paul Lazarsfeld, the director of the Bureau of Applied Social Research at Columbia University and who especially pioneered the use of surveys to psychoanalyze American voting behaviour, and by the 1952 Presidential election along with Madison Avenue advertising agencies were firmly in control of Dwight Eisenhower's campaign, utilising Lazarsfeld's work. The power of television and its hypnotic influence on the electorate became a reality in the 1952 elections. "Batten, Barton, Durstine & Osborne – the fabled 'BBD&O' ad agency – designed Ike's campaign appearances entirely for the TV cameras, and as carefully as Hitler's Nuremberg rallies; one-minute 'spot' advertisements were pioneered to cater to the survey-determined needs of the voters.

"This snowball has not stopped rolling since. The entire development of television and advertising in the 1950's and 1960's was pioneered by men and women who were trained in the Frankfurt School's techniques of mass alienation. Frank Stanton went directly from the Radio Project to become the single most-important leader of modern television."[59]

The whole idea behind this type of 'social brainwashing' is to better understand the electorate's response to policy dictates of the elite. "If you want people to believe something, then all you have to do is get a poll taken that says that it is so and get it publicized, preferably on television." The speaker is Hal Becker of the Future's Group.

We are elite's test cases, to be nudged and prodded in the right directions, all along making us believe that our decisions are based on our solid understanding of the body-politics around us, whereas in reality, we are a 'targeted population groups' whose messages are adjusted to the apparent needs of the masses. The results of the carefully constructed polls, fed back to the population through the mass media, provide the basis for shaping opinion. This process is created at Tavistock and described in one of the Tavistock manuals as "the message reaching the sense organs of persons to be influenced."

It was this juggernaut that transformed most Americans who had never heard of Saddam Hussein and could never dream of showing you Iraq on a map, into a vile demonstration of hatred for an unknown 'enemy', who, we were told, wanted to kill our President and exterminate us as a country. Sound familiar?

How could this be, you ask? Because these images of Saddam, world enemy number one, became an image of evil through the power of television. Just as Hitler and Stalin visually represented evil. How about a little test? Who, in your opinion, is the enemy today? The Chinese? The Korean? Putin? Or is it your fellow man? We have become the enemy, the source of destruction of society. If you believe the evil Freud, "not all men are worthy of love." The sinister environmentalist movement with their anti-human view of man explicitly states this very fact in their degenerate propaganda. "Save the little animals. Kill the people."

Still, we don't seem to get it. Why?

Because the methodology of polling hasn't changed since its inception. Complicated issues are reduced to simple sets of choices. [Saddam Hussein is a bad man. Do you agree?] However, what is this assertion based on? On whose vision? How do you define badness? From the perspective of Iraq's population, the invading U.S. troops were bad and Saddam was good. These types of questions fit directly into the profile of Western Europe and North America: "Keep it simple, stupid."

We are talking about Freudian mass psychology and its appeal to infantilist, animal-like behaviour, the touchy-feelie thing, designed to bypass creative reasoning powers of every individual, informed by moral judgement and the eternal search for universal Truth.

In fact, when it comes to television, the issue of truth has never, in fact, been an issue. Television is not about the 'truth', it is about creating reality. It does not matter one bit whether the images you see on television are real or copied and pasted from past events, because people believe them to be real, immediate and thus true. For example, in the March 2010 earthquake in Japan, mass media showed images of empty supermarket shelves stating brazenly that

Japan was undergoing the worst water rationing since World War II. However, the images of empty shelves were taken from stock photography and had little to do with the earthquake or lack of bottled water. In this way, reality, as conveyed by the nightly news, obliterates Truth every night. Emery and Trist indicated that "the more a person watches television, the less he understands, the more he accepts, the more he becomes dissociated from his own thought process... television is much more magical than any other consumer product because it makes things normal, it packages and homogenises fragmentary aspects of reality. It constructs an acceptable reality (the myth) out of largely unacceptable ingredients. To confront the myth would be to admit that one was ineffective, isolated and incapable... It (television images) becomes and is the truth."

In this manner, the nightly stories on your favorite television news show serve to reinforce your opinion of what the world around you looks and smells like. And what is that opinion? That we live in a world of unspeakable violence and perversion; that human beings are degrading and violent creatures who murder, rape, destroy and spread hatred everywhere they go in the name of whatever –ism happens to be in vogue at the moment.

IDIOTSPEAK

Have you ever paid attention to the level of language used in newscasts? You haven't, have you? In fact, every newscast in the western world follows the same language pattern: simple verbs, oodles of nouns and very few long sentences. Short sentences, simple vocabulary. In other words, sound bites. "President won't run for re-election," says the newsreader, "details in half an hour." This too, is deliberate. Through language, its beauty and complexity, "man communicates the ideas and principles of his culture from one generation to the next."[60] The language of television is Aristotelian, through the mere naming of objects, as in a fixed, counted universe: Man, dog, criminal, president, car bomb, economy, bad, good. "There is no creative thought going on, no attempt to engage the mind, merely to imprint an image in a person's brain."[61] And that's what you call brainwashing. However, brainwashing by television is not an overnight thing, rather it is a cumulative effect over a long period of time that succeeds in changing the paradigm of our society. Don't believe me? Look around and think back a generation.

The language of television news has its roots in linguistic work during World War II as part of H.G. Wells' Ultimate Revolution; the elimination of all national languages in favor of 'Basic English', future world language, which had a vocabulary of only 850 words. This degenerate concept was created by a British linguist, C.K. Ogden. Some people in the highest levels of the British oligarchy, including Churchill, saw the potential brainwashing value in what Ogden had done. By using 'Basic', coupled with the mass media, a large number of people could be given a simple message without complicated thoughts getting in the way.

In Wells' future world: "The English most speak and write today is very different from the English of Shakespeare, Addison, Bunyan or Shaw: it has shed the last traces of such archaic elaborations as the subjunctive mood."

However, human powers of hypothesis and creative reason become a force for improvements in the order of nature, and thus cannot be encased into an 850 word vocabulary. Through languages, nation states were given meaning and through them our quest for Truth. In ancient Greece, in Plato's Meno dialogue, "Socrates brings out the innate genius in the slave boy, by encouraging him to solve the problem of doubling the square. He thus proves that slavery is unjust, by showing the creative nature of the human species. Plato's pro-imperial opponent Aristotle, assuming that man has no soul, declared that nature has destined some to be slaves, and has made others their masters."[62]

Tavistock knows that ideas are more powerful than any standing army. To get people to accept their imperial ideas, Tavistock seeks to control the way people think, especially science, the area where human powers of hypothesis and creative reason become a force for improvements in the order of nature. If you can control how we think, you control how we respond to events around us, no matter what these events might be. This process is called 'paradigm shift', an overturning of existing sets of assumptions about society.

Historically, the fight to preserve language as the cornerstone of nation-state republics goes back to Italian playwright, Dante Alighieri.

As sociologist Dennis Speed writes on the pages of Fidelio Magazine, "It is important to identify the fact that the people of the Italian Renaissance spoke a language that did not exist two hundred years earlier, but was invented by the poet Dante Alighieri. Dante, who lived from 1250 to the early 1300s', struggled to invent a language that could resurrect the most profound ideas of human thought even if his own era should ultimately commit suicide. This suicide, in fact, did occur through the banking collapse caused by the Bardi and Peruzzi families.

"Without Dante's gift of the Italian language, which was shaped by him from over a thousand local dialects, there could not have occurred the Renaissance. Dante's *Divine Comedy* refined the canto form of sung poetry. His followers, Francesco Petrarca, sought to advance language further with the invention of the soneto or sonnet. But, Petrarca's friend Giovanni Boccaccion was assigned by Petrarca to a different project, called the Decameron. The Decameron was written in order to prevent the whole of Italian society from sinking into cultural pessimism and dying out during the Black Death of the 1340's and later. It consists of satirical stories, many quite risqué, which recount the tragedy of Europe's suicide in a way designed to make people laugh at themselves and repeat the stories, and hopefully not the behaviour which had destroyed them. Thus, they would learn Dante's Italian.

"Geoffrey Chaucer, attended a lecture by Boccaccio on Dante in 1375, and got the idea to do the same thing in English. Thus was born *The Canterbury Tales*, which recount the hilarious tales told on a religious pilgrimage to the church of Canterbury. The English, being every bit as licentious as the Italians, also repeated the stories and thus learned how to speak English. Later, Shakespeare imported the sonnet of Petrarca into English. Your literate English, is in large measure Italian, a sort of grandson of Dante's Italian. Later, the Christian humanist figure Erasmus of Rotterdam inspired his student, Francois Rabelais, to do for France what was done by Boccaccio for Italy and Chaucer for English. Thus was born the astounding character Gargantua and with it, the French language. And in Spanish, we have the great example of Don Quixote."[63]

In this way, languages, created by poets, lifted the populations that had been dominated by ignorance and thereby ruled over. In fact, the languages created the nations, not the other way around.

WAR ACROSS GENERATIONAL DIVIDE

People in their sixties are called the Baby-Boom generation. The values common to the Baby Boomer generation are synonymous with the paradigm shift in society beginning with the 1960s. That policy change is de-industrialization of America. De-industrialization society is a no-future society because any self-respecting nation-state can only be truly independent if it can provide welfare for its citizens. How were the people who run the world able to impose this de-industrialization on society? Think about the profile of the American population.

The generation with the strongest set of moral values in American history is the generation that fought in World War II. They were born before the advent of television. They were the hardest generation to brainwash. Their children, the baby boomers, became the targets of the brainwashers. In fact, with the advent of television, the baby boomers have spent their entire lives being brainwashed. Think back to the goal of brainwashers: to make each succeeding generation more infantile, more animal-like, more amoral, thus easier to control. The world we are part of today is more conventional, less concerned, more sensationalist and more spiteful. The impetus of modern thought has been towards reducing the sphere of individual moral responsibility. Human behavior is seen increasingly as the product of impersonal forces.

And how could it be different? Our parents were brought up on a "healthy" dose of television diet. We were also reared by television. Our children are educated through mainstream TV. In other words, three successive generations subjected to television brainwashing, have no conscious memory of anything different.

Think about it. People have never quite digested television. The mystique which should grow stale grows stronger. Our moral compass has shifted to such an extent that we make celebrities not only of the men who cause the events but of the men who act them out on the screen, who lipsync old songs for hand-picked studio audience and of those out-of-work actors who sell cheap trinkets, yogurts and deodorants during commercial breaks of old film re-runs.

And once our moral compass shifted something happened to us and with us. The TV set became the moralizer. It began telling people what to do. Suddenly, the behavior of the children – the drugs, the sex, the anti-social behavior doesn't seem as shocking as it once did. It became easier to rationalize it, to explain it away with the help of the messages contained in television programming. "And when the adults are infantile already, it is more easy to accept the infantilism in their children."

CULTURAL WARFARE

Now, let's have a look at the people who create the idea content of the brainwashing. Most of them are in the approximately 35-45 year old range. In other words, the programmers themselves were weaned on television since birth! So, don't look for a scenario where three 90-year-old geezers sit in a dark room, holding hands, staring at a crystal ball, planning the world's future. It doesn't work that way. It is more like a group of infantile jerks in sneakers and goofy haircuts, sitting around in an air-conditioned room overlooking the Big Apple, kicking around ideas for next year's major shows.

Once they come up with the ideas, they pass it on to another group of infantile jerks called scriptwriters who will put these ideas into simple dialogues full of nouns and simple verbs. That product will then be placed in the hands of producers and directors for actual production. This is done the world over, from the United States to Canada and Western Europe. Even thousand-year-old oriental cultures, such as the Japanese, have their own versions of all major and minor U.S. hit situation comedies, adapted to their own market.

Once the shows are designed and the scripts written, the actual brainwashing messages are inserted after the fact. The jerky screenwriters and brainstormers rely on "experts" from Tavistock and Frankfurt School who work as "advisers" on most programs. For example, "there can be no discussion of homosexuals without first deferring to the 'Gay Rights Task Force' who make sure that the issue is being dealt with 'sensitively.' Similarly, all children shows and that includes all Disney programming employ child psychiatrists who help shape content." Furthermore, the entire environmental idiocy is scripted into television shows, films and into the news.

Take, for example, a 2008 Hollywood science fiction thriller, *The Happening*, starring Mark Wahlberg, that "follows a man, his wife, and his best friend along with his friend's daughter, as they try to escape from an inexplicable natural disaster. The plot revolves around a mysterious neurotoxin that causes anyone exposed to it to commit suicide."[64] The message is easily discernible. People are the enemy of nature and Planet Earth is rebelling against us, ruthlessly killing the enemy.

Those who protest and are opposed to such idiocy are shunned and scorned by their co-workers, friends and family. If you weren't completely won over, you became tolerant or at the very least, you kept your mouth shut as the lonely voice of reason. Isn´t that right?

The New York-Hollywood social community of 'creative' people function in what brainwashers call a leaderless group: "They are unaware of the real outside forces that control them, especially unaware of their own brainwashing by 30-40 years television viewing. They believe themselves free to create, but they can lawfully only produce banalities. Ultimately, these creators of our television programming turn to their own brainwashing experience and values for their 'creative inspiration.'"[65] When one producer was asked how he determined what was in his shows, he replied, "I think of myself as the audience. If it pleases me – I always think that it is going to please the audience." Indeed it does. How many people who watch television are aware that what they see on the boob tube reflects the morals and conscience of the people who make them happen?

Let's have a look at the two most 'popular' television shows in Spain to have garnered as much attention in the past decade as *Operacion Triunfo* y *Gran Hermano* [Big Brother]. Why are they so popular? One might suspect the content itself – a startling new idea, but the reason might also be social acceptablity, or even just plain hype. Each of these programs, with their claims of righteous generosity, even-handedness and ratings measured fun, contains no new arguments and presents no compelling reasons why the studio audience, the show's participants and all those simple folks watching at home shouldn't be locked up in perpetuity.

By any standard, *Operacion Triunfo* is an impressive achievement, a cultural prototype for the gullible and the sick. Their participants are better looking, slicker, more calculating, more diabolical, more astute in optimum crowd manipulation than their predecessors – a triumphantly bizarre television 'celebrity' moment of puerile distortionism.

Television, in fact, has given status to the "celebrity" which few real men attain. And what is a "celebrity" if not the ultimate human pseudo-event, fabricated on purpose to satisfy our exaggerated expectations of human greatness. This is the ultimate success story of the twenty first century and its pursuit of illusion. A new mold of human emptiness has been made, so that marketable human models – modern 'heroes', could be mass-produced, to satisfy the market, and without any hitches.

The world we have been peering into is somehow beyond good and evil. It is a world of sentimentality, of makeovers, of people who are willing to shed a tear just before a commercial break, and then return with uplifting visions of a life with family by show's end.

It is a world of post baby boom, post paradigm shift and post reality. It is a world of us versus them, of our parents generation against our generation. And the divide could not have been greater than it is today. "Bridging the two generational value structures, was the television set in your living room. And what did it tell you: to compromise, to learn to speak to each other. It provided you with solace, as your world collapsed and changed."[66]

Then, there is Big Brother. There are those who argue that each show is a small morality play, with the audience booing the villain and supporting the good guests. The show insists on as much – the tone, a throwback to the time-tested model of the past, is almost always conspiratorial. The nincompoops are more self-righteous, more vehement, and more paranoid. Instead of saying indefensible things and trusting that the viewer will love him anyway, they explain their hardships and plead their case, which is a very unreasonable thing to do with an abjectly stupid audience. I find it disturbing to see people allowing themselves to be used like Kleenex for shallow ends. Initial triumphalism, giving way to grudgy defeatism – is today's world as seen through the prismatic binoculars of an intrusively globalized, theraupeutized and Coca-Cola-ized American morality.

"Attacking the entertainment of a therapeutic culture is a way of attacking its values: publicity over achievement, revelation over restraint, honesty over decency, victimhood over personal responsibility, confrontation over civility, psychology over morality."[67] According to a well-informed source at TVE, the more intellectual you make the shows, the more the viewers yawn and go away. But then, we already know that. And why should we be surprised? In this society, shame is fame and sin is an instrument of upward mobility. And upward mobility is the modern way.

In *Webster's English Dictionary*, fame is defined as "the frenzy of renown." The frenzied, meaning people who are intoxicated by synthetic significance – are complicit in this farce. Forty years ago, in his book *The Image*, Daniel Boorstin argued that the graphic revolution in television had severed fame from greatness which generally required a gestation period in which great deeds were performed. This severance hastened the decay of fame into mere notoriety, which is very plastic and very perishable.

The doctrine of "celebrity" triumphalism is inherently inclusive; it proceeds from the assumption that everyone already loves the wealthy and the pseudo-famous. Most pseudo "celebrities" seems to have understood that their life is a constant conjuring trick. There is an incurable precariousness to their position as they try to live off derivative dignity from an anachronistic concept while cultivating the royalism of a democratic age – celebrity.

That assumption is what makes the celebrity circus so creepy. The pseudo-famous know that they owe their careers to the celebrity industry, know that their tabloid sensibility is well suited to a tabloid culture, know that the whole enterprise is built on bad faith. Still, the ratings continue to climb... .

We have always had a voyeuristic streak. Ours has always been a culture of gawkers. We went to see the bearded lady and the albino man when the circus came to town. But now we want to be a part of the circus – the bearded lady and the albino man rolled into one. And it may be just the right thing for the current moment, in which the sleazy patter of tabloid culture has given way to a jittery onslaughts of bearded midgets and their six-foot tall offspring and guaranteed to appeal to the intellectually challenged (reason being highly optional). The abnormal is normal, and deviancy will be back right after the commercial.

Disturbing as I find the anachronism of the Variety shows, I am even more distressed by their pervasive disingenuousness. The hosts and the guests alike omit facts, misuse words, and seem unwilling to admit the consequences of their own actions. I have never seen anything more grotesquely inadequate in a way of confession.

As Maureen Dowd writes in the *New York Times*, "Confession began in a small dark box, with a screen separating priest and penitent. It is still in a small dark box, and there is still a screen, but this one is in everyone's living room. In the new medium of confession, people violate their own privacy, spilling their guts not for absolution but for syndication."[68]

In fact, we are living in a world where conspicuousness passes for distinction, and the society column has become the roll of fame...

André Malraux believed that the third millennium must be the age of religion. I would say rather that it must be the age in which we finally grow out of our need for religion. But to cease to believe in our gods is not the same thing as commencing to believe in nothing. To believe, we must take on the richness of a man, his existential density, immortality, eternity and not the sectarian, simplistic, visceral millenarianism. Memory saves people from oblivion. The inherent danger, however, is that the prerequisite is lacking: curiosity deriving from respect for deeply alien cultures.

That is not the main lesson of what is happening today, but it is a crucial lesson. The public debate increasingly is in the hands of the mentally retarded. This degeneration has played a large role in a paradigm shift of our society. What is that shift, again? Man is to be told that he is a beast, so that he might be controlled like one. The brainwashers of Tavistock adopted a bestial view that the human mind is simply a blank slate, which avoids pain and seeks pleasure. "It was from this standpoint that the Tavistock Institute developed its peculiar techniques for creating a mass psychology."[69]

TELEVISION'S HIDDEN MESSAGES

In a 1944 report, Theodore Adorno, one of Frankfurt School's premier thinkers, postulated that media such as radio and especially television could be used to make people "forcibly retarded." Listen to his words: "Television aims at the synthesis of radio and film, and is held up only because the interested parties have not yet reached agreement, but its consequences will be quite enormous and promise to intensify the impoverishment of asethetic matter so drastically..." Twelve years later, Adorno wrote that "television is a medium of undreamed psychological control." In his ground-breaking essay "Television and the Patterns of Mass Culture," Adorno writes, "...

investigative systematically socio-psychological stimuli typical of televised material on both the descriptive and psychodynamic levels, to analyse their presuppositions, as well as their total pattern, and to evaluate the effect they are likely to produce. This procedure may ultimately bring forth a number of recommendations on how to deal with these stimuli to produce the most desirable effect..."[70] Adorno states that "all television programming contains overt messages as defined by the plot, characters, etc and a hidden message," far less obvious. These hidden messages in themselves are the brainwashing content, while "the overt message, such as the plot are the carriers of that brainwashing content."[71]

Here is a bit on Lonnie Wolfe's write up of Adorno's message: "*Our Miss Brooks*, a popular situation comedy pitted a trained professional, a school teacher, against her boss, the principal. Most of the humour, according to Adorno, was derived from situations in which the under-paid teacher tried to hustle a meal from her friends. Adorno 'decodes' the hidden message as follows. 'If you are humorous, good-natured, quick-witted, and charming as Miss Brooks is, do not worry about being paid a starving wage. You can cope with your frustration in a humorous way and your superior wit and cleverness put you not only above material privations, but also above the rest of mankind.' This message will be called forth years later, as the economy collapses in the form of a "cynical anti-materialism." It came forth with a vengeance among the 1960s 'lost generation' and the first wave of counterculture."[72] If we are to extrapolate this, the more confusing life becomes, the "more people cling desperately to clichés to bring order to the otherwise un-understandable," Adorno writes.

In one of his other reflections, Adorno predicted that the "creative sissy will find an important place in society." Several shows featured characters who were artistic, sensitive and effeminate males. "Such images cohered with Freudian notions that artistic creativity stemmed from either a repressed or actual homosexual passion. These effeminate, sensitive males usually come up against the other more, aggressive male 'macho' images, such as cowboys, who are uncreative."[73]

ANIMAL KINGDOM AND US

Another key area of brainwashing is the creating of identity between man and animal. From children's cartoons to television shows and major motion pictures portrayed animals acting like they were human beings. Over time, according to studies, kids lost their ability to differentiate between most animals and human life. Format shows such as *Lassie*, the animal was a 'hero' who often defeated single-handedly 'bad' people who wanted to hurt Lassie's child friends. "All of this identification with the animal, and the blurring of the distinction between what is human and what is animal played back a generation later in the lunacy of the environmental movement."[74]

But even before there was television, another mass media phenomenon preconditioned young people for the non-rational, audio-visual experience: the Walt Disney full-length feature cartoons such as *Snow White, Cinderella, Sleeping Beauty, Pinocchio*, and more recently *Little Mermaid* and *Beauty and the Beast*. Without ever realizing it, adults and children alike have been subjected to over 60 years of some of the most vile and pernicious propaganda in modern history. "These feature length cartoons were aimed at becoming universal experiences for generations of children and their parents, containing moral messages that would stay with a child through most of his adult life."[75]

What few people realize is that both Walt Disney and his brother were producing WWII propaganda films, managed by the Committee for Morale. "Disney cartoons were not to make people think, but to feel, something that Disney said, "would unify his audience of parents and children at an emotionally infantile level."[76] Disney and the degenerates at Frankfurt School and Tavistock are on the same page here, with people such as Adorno who spoke of using the media and its power to convey emotion-laden images to force a retardation of adult society. Thus pictures and sound were to create a mental image that would seem real, to give his characters an emotional dimension. In such a dream-like state, people were more willing to accept the vivid imagery of his cartoons as real. Listen to Disney and if you are a human being instead of a two-legged animal, his words should make you sick to your stomach: If all the world thought and acted like children, we'd never have any trouble. The pity is that even kids have to grow up.

Without realizing it, you are being bombarded with an ultra-heavy dose of some of the vilest Jungian symbolism through the mass entertainment media, "creating mythological worlds of 'superheroes' and 'supervillains,' while introducing character representation of archtypes such as the 'Great Mother' or the 'Wise Old Man' or the 'Maiden' or the 'Eternal Youth.' This was no coincidence as the largest concentration of Jungians in the United States continues to be in Hollywood where numbers of producers, directors, screenwriters, actors and actresses underwent Jungian 'dream therapy.'

"In addition, musicians, lyricists and other components of the rock scene, influenced by the LSD experience, gravitated toward Jungian thought, and inserted his symbolism into their songs. No where are there sharper representations of Jungian archtypes than Disney cartoons. His cartoon representations were totally coherent with Jungian concepts, especially his consistent presentation of morality in the absence of the teachings of the Judeo-Christian concept of the Good. To the extent that good triumphs, it does so only through the intervention of magical powers of fairies which are more powerful than evil. This is the essence of the kind of spirituality which Jung speaks of – the symbolic fight between 'dark' and 'light' forces, outside the control of human reason."[77]

But there is a scarier aspect of the entire Disney circus, with its grizzly dip into the macabre. For the past 60 years, one of the most popular children's shows has been the Mickey Mouse Club with its synthetic mix of real people, live music, cartoons and live interaction between humans and people dressed up as animal characters. How many people realize that the Mickey Mouse Club was a sinister experiment in mass brainwashing of children through television? Each child at home was 'indoctrinated' in a membership ritual, with prompting from the television, and was urged to singalong with words flashed on the screen and chant things as instructed by their television group leader. They did so while wearing their 'mouse ears,' which were designed to make them identify with the animal figure, Mickey Mouse. At the end of the show, there was a sermon by a group leader, a young male adult, whose preaching was reinforced from the in-studio Mouseketeers. All of this done while children at home and on the stage wore their ears and gave their 'club salute.' How many people realize that while they were giving their Mouseketeer salute, they were actually being fed a new pagan-like religion, and its newly installed god, the mouse?

If you extrapolate it a bit further, parents stepped aside and allowed a mouse, or rather television, through a Mickey Mouse, to give values to a generation of children who would teach the same values, in most cases to their children and their children's children. I repeat, you are looking at three successive generations subjected to television brainwashing, without any conscious memory of anything different. In another time and in another European country, another generation of children was given values to them in an organized form by someone other than their parents. "The Hitler Youth of Nazi Germany. They too had their rituals, their uniforms and symbols, and their songs. They too had their leaders, who preached sermons. They too were told to listen to their parents and be patriotic, polite and well-behaved."[78] The trick was to make the Nazis disappear, but not their ideals. Nazis yes, but without its modicum. Mickey Mouse and Hitler. Can you see a parallel?

THE INVISIBLE EMPIRE

With all this talk about Dante and Disney, Mickey Mouse and paradigm shift, Judeo-Christian culture and Aristotle, Freud and Jung, Tavistock and Frankfurt School, you might have forgotten the main theme of this chapter. Let's have Eduard Bernays, Freud's nephew, tell it like it is: "The conscious and intelligent manipulation of the organized habits and the masses is an important element in democratic society. Those who manipulate this unseen mechanism of society constitute an invisible government which is the true ruling power in this country... we are governed, our minds are moulded, our tastes are formed, our ideas suggested largely by men we have never heard of... Our invisible governors are, in many cases, unaware of the identity of their fellow members of the inner cabinet... Whatever attitude one chooses to take toward this condition, it remains a fact that in almost every act of our daily lives, whether in the sphere of politics or business, in our social conduct or ethical thinking, we are dominated by a relatively small number of persons...who understand the mental processes and social practices of the masses. It is they who pull the wires which control the public mind, who harness the social forces and contrive new ways to bind and guide the world."[79] Can anyone in their right mind, still deny it? Can anyone deny that there has been a mind-boggling paradigm shift in our moral values in the past two generations? Look around you! Someone is doing this from behind the scenes – or is it still not clear!

Now that we have settled the question of Mickey Mouse, let's have a look at another all-time children's favorite – Sesame Street. The goal of the show conceived in 1966 was "to master the addictive qualities of television and do something good with them, such as helping young children prepare for school."[80] The first thing a child sees is a show dominated by animal-like creatures with human characteristics, the famous Muppets. Before we get into the brainwashing aspect of it, I would like you to listen to one of the Tavistock poster boys – Malcolm Gladwell. In his own words: "Sesame Street was built around a single, breakthrough insight: that if you can hold the attention of children, you can educate them."[81] Sesame Street, according to Wikipedia, "was the first children's show that structured each episode and made, as Gladwell put it, 'small but critical adjustments'[82] to each segment to capture children's attention. After Sesame Street's first season, its critics forced its producers and researchers to address affective goals more overtly. The affective goals they addressed were social competence, tolerance of diversity, and non-aggressive ways of resolving conflict, environmentalism, which was depicted through interpersonal disputes among its residents.'"[83]

That's one of the hidden messages. Do you see? *Sesame Street* preaches its personal brand of degeneracy by having beastly Muppets openly moralize about the environmental issues, social competence, tolerance and more. Another hidden message is that the 'only' correct solution, no matter the problem, was compromise, to learn toleration. All of this is done while the Muppets teach the children to read and write. But there can never be compromise with evil, for those who hold universal Truth as their ultimate objective in life. That's part of moral rectitude and character that defined all great nations and people of ideals.

"Shortly after creating Sesame Street, its producers began to develop what came to be called "the CTW model," a system of planning, production, and evaluation that did not fully emerge until the end of the show's first season.[84] The CTW model consisted of four parts: 'the interaction of receptive television producers and child science experts, the creation of a specific and age-appropriate curriculum, research to shape the program directly, and independent measurement of viewers' learning.'"[85]

But this is a big lie. Studies show that the show does not enhance learning; "in many cases, it appears to inhibit their ability to understand more complicated ideas. More importantly, the studies indicate that the children appear 'addicted' to the show, and by that 'addiction' to become addicted to television viewing in general."[86] As Neil Postman, a New York University professor wrote in his book *Amusing Ourselves to Death*, "If we are to blame Sesame Street for anything, it is the pretence that it is an ally of the classroom...Sesame Street does not encourage children to love school or anything about school. It encourages children to love television."

There is absolutely no question that Sesame Street is about as Establishment as you can get. Its money came from the Rockefeller controlled Carnegie Foundation and Ford Foundation who have pushed the Sesame product into a global network. Ford Foundation money comes from the Central Intelligence Agency, hardly a paragon of child education; David Rockefeller-controlled Trilateral Commission, Council on Foreign Relations, Carlyle Group whose members include former President George W. Bush, former Secretary of Defense Donald Rumsfeld, not to mention some of the lesser known members of the bin Laden family and Skull & Bones Yale University secret society. You didn't know that, did you? The David Rockefeller-controlled Carnegie Foundation is one of the key insiders in the powerful Bilderberg Group who work diligently from behind the scenes to debase the world's education in order to bring us down to the level of culled human cattle.

It is run as a business, and has always been so. It has never lost money and continues to rake in substantial profits, all in the name of child-based education. Children weaned on this garbage have helped make Sesame Street into a $1.5 billion industry, one that unlike the rest of the economy is expanding each year.

You don't agree? Let me ask you a question. Don´t you consider it odd that your young daughter wants to grow up and be just like Miss Piggy or Big Bird?

THE MEDIA ERASER

What about adult brainwashing? How do they get to us, the 'big people'? One of the all time favorite Spanish television shows is *Cuéntame Que Pasó* (Tell Me What Happened). The show premiered on Spain's public television on 13 September 2001. The story begins in April 1968 and paints a day-to-day portrait of Franco's Spain through the adventures of a middle-class Alcantara of the 1960s, then the 1970s and into a modern democracy

So far, so good. However, does the show focus on the real horrors of the period? The drugs, the destruction, the collapse of social values, the changing paradigm of society? Not at all. Instead, we are treated to a daily diet of superficial boy loves girl thing, with television parents never succumbing to stress and cliché family quarrels. Franco's death and subsequent changes are treated with a kind of degrading moralism.

Cuéntame Que Pasó is what's called 'controlled flashbacks' for the baby boomers "to what they would now like to think the 1960s were like. By doing so, the producers have put you in touch with the most infantile and banal emotions, and made you feel nostalgic for them. The hidden message: in these difficult times, one had best cherish cling to memories and values of one's infantile past. Think of television as a big eraser, wiping away your real memories of the past, the reality of the way things really were."[87] The tube superimposed a twisted and distorted view of that reality through appeal, not to your mind, but to your infantile emotions. The idea of rewriting history through the use of media is not new. But this is the first time that it is being done while the people who lived the history are still very much alive.

THE WORLD OF ADVERTISING

Television is the most effective way to brainwash you, but it is not the only way. Madison Avenue in New York, the epicenter of advertising, has also contributed their fair share, and by extension the 'cult of celebrity,' would play a role in disseminating what Tavistock wanted you to believe. In the 1920s, Edward Bernays took peacetime propaganda and renamed it public relations. "Public Relations created the consumer society, which was said to be the cherry on the icing of the free market capitalism cake, with the introduction of celebrity endorsements and product placements."[88] PR consolidated the theories of mass psychology and schemes of corporate and political persuasion upon the average person's beliefs by appealing 'above the mind,' directly to emotions and instincts.

"For more than 55 years now, people have been watching ads which, through the clever use of music and image, have attempted to manipulate subconscious drives and instincts to sell products. Most run less than a minute, but contain numerous images and quite often a catchy jingle."[89] The formula is simple. If you are a young person in western culture, advertising meets your information needs about growing up and living life. If you analyze it, there is virtually no difference between a non-reasoning emotional appeal to watching television shows and watching television advertising. Both sell a point of view. No lectures. No lessons. No right way or wrong way. It simply expresses, through the life experiences of others, what life is or can be like for the target demographics. It's very post-modern and very deceptive. Reality works, because it can be hyped as such, and manipulated.

One point of view could easily be nostalgia. Nostalgia, as in *Quéntame Que Pasó*, is an interesting thing, acting as a common past that binds those of a particular time or generation. When used correctly, nostalgia can also be very handy in manipulating people. Advertisers discovered this long ago, which may be one reason certain fashions and trends from the past three decades have made a comeback in the last ten years.

For what appears in advertising today creative, resourceful and unscrupulous people constantly try to discover what others value most – then look for some way to attach their product to the stars.

By their very nature, no products can help us attain the ideals that are visually promised, such as family togetherness, personal power, self-esteem, sociability, security, sex appeal and a clear orientation in an evermore confusing world.

Advertising is the process of manufacturing glamour. The state of being envied is what constitutes glamour. Advertising, then, is about the solitary happiness that comes from being envied by others. But envy has a dark side which largely has been lost to twentieth century thought. Since medieval times, envy has been considered the major term for identifying the causes of human suffering. Like despair, envy derives from the separation of the person from the object of desire, combined with a sense that one is powerless to attain what is desired. In envy, the urge to reach out becomes the urge to destroy.

And yet, at the heart of this hatred, lies the remarkable depth and simplicity of human longing – a longing for life, ideals, values, vitality and love. A longing for connection. A longing for beauty. It is a longing that projects itself optimistically through symbols, images and idealized concepts.

Advertising is the consumer culture's version of mythology. No society exists without some form of myth. Thus, it is not very surprising that a society which is based on the economy of mass production and mass consumption will evolve its own myth in the form of a commercial. Like myth, it touches upon every facet of life, and as a myth it makes use of the fabulous in its application to the mundane.

Why would the makers of commercial want to evoke hate and envy? In his book, *Quest for Mind: Piaget, Levi-Strauss and the Structuralist Movement*, comparing Piaget and Levi-Strauss, Howard Gardner wrote: "Myths are designed to deal with problems of human existence which seem insoluble; they embody and express such dilemmas in a coherently structured form, and so serve to render them intelligible. Through their structural similarity to given 'real world' situations, myths establish a point of equilibrium at which men can come to grips with the crucial components of the problem. Thus a myth is both intellectually satisfying and socially solidifying."

Commercials driven by value-laden images which are unrelated to the product may be alienating us from the very values they exploit, confusing us about how to attain these values, laying the groundwork for despair, resentment and apathy.

Because products do not provide the kind of psychic payoff promised by the imagery of advertising, we are left to doubt whether anything can. If we follow this doubt, we wind up contemplating a state of mind in which a black hole surrounds almost every product like a ghostly negative of its radiance – the black hole of failed promise.

And into this black hole, by advertising's exploitation of so many ideal images, steps any religion that promises to cut through the cycle of idolatry and connect us with the one great ideal that trascends all others: God, immortality, cosmic foreverness, spiritualism, etc. Thus, in many respects, advertising has co-opted religion for its own purposes.

That is a disturbing possibility; but another possibility is even more disturbing. Could we be producing a generation that distrusts ideals altogether, because most powerful, convincing and forceful presentations of those ideals occur on television commercials – where the ideals are prostituted in the service of sales? Are we responsible for the creation of the most disillusioned generation in the history of mankind? A generation that will have difficulty not hating beauty of the kind used to manipulate and disappoint them in advertising? Worse still, will they also hate to be delicately overpowered by real beauty when they encounter it in the world? Will people continue to be able to hope, have faith, set goals, and believe in something beyond themselves?

Beauty has not always seemed so complicated. "From the times of the Greeks to the early 20th century, philosophers and poets connected beauty with such glorious ideas as truth and harmony. For Dante, such ideals were the guiding lights that illuminated existence. One has only to remember the conclusion of Keats' "Ode on a Grecian Urn:"

> Beauty is truth, truth beauty, that is all
> Ye know on earth, and all ye need to know."[90]

The magazine cover in which feminine seductiveness, or the comic strip in which narrative clarity has absorbed all the artist's creative impulses, are contemptible, not because of their insistence on sex appeal or their concentration on anecdote – Titian and Boucher could rival them in the former, Giotto and Goya in the latter, but because of their lack of enrichment from within. They are hollow for the same reason that abstract artists are small and often insignificant.

The success of modern advertising reflects a culture that has itself chosen illusion over reality. The supernormal images of perfection presented on the media are worth some thought, because any kind of a guiding image has a double nature.

But idealized images are uplifting only when there is some way to move from where you are in the direction of the values implicit in the image. If there is nothing to connect us with the image, so the ideal seems unattainable, we feel cut off from it, thus promoting not the joining of the audience and the ideal, but just such a separation.

Advertising promotes despair first by surrounding us with images of unattainable perfection. Second, advertising promotes despair by implying that the product will deliver the ideal – when it can´t. People do not need new cars every three years, plasma television brings little enrichment of human experience, the length of a dress does not affect the expansion of knowledge, or increase the capacity to love. As one critic of advertising put it: "Sadness betrays the idyll (of advertising's more-than-perfect world)."

Despair is a natural by-product of the experience structured into the way advertising promises to deliver the values implicit in its hypernormal images.

As Vladimir Nabokov accutely remarked, "In our love for the useful, for the material goods of life, we have become easy victims of the advertising business. The rich philistinism emanating from advertisements is due not to their exaggerating or inventing the glory of this or that serviceable article but to suggesting that the acme of human happiness is purchasable and that its purchase somehow ennobles the purchaser. The amusing part, of course, is not that it is a world where nothing spiritual remains except the ecstatic smiles of people serving celestial cereals but that it is a kind of satellite shadow world in the actual existence of which neither sellers nor buyers really believe in their heart."

*****

"Being immersed in Classical culture," writes Harley Schlanger, "develops your understanding of universal history, of the ideas behind the battles, to prepare you to act as a leader in your own time. And at the heart of truly great Classical culture – as in tragedy – the author demonstrates that a tragic destiny is not inevitable, but that there is a path of action by which tragedy may be averted."[91]

As illustrated by the facts in the case of the famous Brotherhood of the Common Life, the student who benefitted from a classical education approach of cognitive processes is the student who at the very least has a decent chance of becoming a qualified genius.

This quality of cognition is what separates us from animals. These are the true discoveries of principle, the discoveries made by individual minds, that have permitted the species to increase the potential population density of humanity, from the level of a higher ape of a few million strong to over three hundred million in the fifteenth century, and over seven billion today.

History, if taught from the perspective of cognitive development of whole cultures, and of the individual within the culture, also acquires the status of a true science. It is through the valid ideas, that humanity is above the apes, and most of humanity rises above the barbarous and feudal condition of "human cattle." History, in this sense, is the history of ideas, production, circulation, and its realization.

However, the first prerequisite to understanding our rightful place in the Universe starts with you, reader. Go on, turn off your television and most importantly, keep it off.

Endnotes

1. Television: Opiate of the Masses – Mental Environment, http://www.familyresource.com/lifestyles/mental-environment/television-opiate-of

2. Ibid

3. Who owns your culture?, Harley Schlanger, *Fidelio*, Vol. XII, No. I summer 2003

4. Turn off your TV, Lonnie Wolfe, *New Federalist*, p.6, 1997

5. Turn off your TV, Lonnie Wolfe, *New Federalist*, p.5, 1997

6. On the Subject of Economy: The Issue Is Humanity, Lyndon LaRouche, *EIR*, April 9, 2010

7. Turn off your TV, Lonnie Wolfe, *New Federalist*, p.7, 1997

8. Ibid

9. Turn off your TV, Lonnie Wolfe, *New Federalist*, p.6, 1997

10. *The Psychology of the Crowd*, Gustave LeBon, reprinted Transaction Publishers, 1995

11. Turn off your TV, Lonnie Wolfe, *New Federalist*, p.7, 1997

12. *Future of an Ilusion*, Sigmund Freud, Horace Liveright, 1928

13. Turn off your TV, Lonnie Wolfe, *New Federalist*, p.8, 1997

14. Turn off your TV, Lonnie Wolfe, *New Federalist*, p.8, 1997

15. Ibid

16. Ibid

17. Eric Trist "The Formative Years, The Founding Tradition, Pre-War Antecedents"

18. Anthony Stevens, *Jung: A very short introduction*, Osford University Press, 1994

19. CV X, para. 354

20. SW , para.388

21. Turn off your TV, Lonnie Wolfe, *New Federalist*, p.9, 1997

22. Ibid

23. Turn off your TV, Lonnie Wolfe, *New Federalist*, p.10, 1997

24. *The Informed Heart*, Bruno Bettleheim, Avon Books, 1960

25. *Unholy Alliance : A History of Nazi involvement with the Occult*, Peter Levenda, Bloomsbury Academic, 2002

26. *Le matin des magiciens*, Jacques Pauwels, Louis Bergier, 1960

27. Ibid

28. *Star Wars and Littleton*, Lyndon LaRouche, June 11, 1999, EIR

29. Theodor W. Adorno et al., *The Authoritarian Personality*, New York: Harper, 1 950.

30. K. Lewin (1942), "Time Perspective and Morale," in G. Watson, ed., *Civilian Morale*, second yearbook of the SPSSL, Boston: Houghton Mifflin.

31. Turn off your TV, Lonnie Wolfe, *New Federalist*, p.12-13, 1997

32. Ibid

33. Ibid

34. Ibid

35. Ibid

36. Ibid

37. George Orwell, *1984*, Signet 1961, 270p

38. Turn off your TV, Lonnie Wolfe, *New Federalist*, p.13, 1997

39. Turn off your TV, Lonnie Wolfe, New *New Federalist*, p.14, 1997

40.   http://www.census.gov/newsroom/releases/archives/facts_for_features_special_editions/cb08-ffse03.html

41. www.pbs.org

42. John Chancellor with Walter R. Mears, *The New News Business*, New York: HarperPerennial, 1995

43. Jeffrey Steinberg, The Cartelization of the news Industry, *The American Almanac*, May 5, 1997

44. Who is behind Wikileaks, Michel Chossudovsky, www.globalresearch.ca, December 14, 2010

45. America's Strategic Repression of the 'Arab Awakening' Part 2, Andrew Gavin Marshall, globalresearch.ca, February 9, 2011

46. *Aid & Abet*, Vol. 2, No.2, pg.7

47. Laughland, John, Fill Full the Mouth of Famine, *Scoop Independent News*, July 29, 2004

48. Jeffrey Steinberg, The Cartelization of the news Industry, *The American Almanac*, May 5, 1997

49. Turn off your TV, Lonnie Wolfe, *New Federalist*, p.34, 1997

50. Ibid

51. Ibid

52. A Wiki for Whistle-Blowers, By Tracy Samantha Schmidt/Washington Monday, *Time* magazine, Jan. 22, 2007 http://www.time.com/time/nation/article/0,8599,1581189,00.html#ixzz1ExtfiKg4

53. http://mirror.wikileaks.info/

54. A Wiki for Whistle-Blowers, By Tracy Samantha Schmidt/Washington Monday, *Time* magazine, Jan. 22, 2007

http://www.time.com/time/nation/article/0,8599,1581189,00.html#ixzz1ExtfiKg4

55. Edward Bernays, *Propaganda*, 1928, reprint, Ig Publshing, 2004

56. The Conspirators' Hierarchy: The Committee of 300, John Coleman, *WIR*, August 1997

57. Michael Minnicino, Frankfurt School and 'Political Correctness', *Fidelio*, Winter 1992

58. The major pollsters of today – A.C. Neilsen, George Gallup – started in the mid-1930's. Another important pollster is Yankelovich, Skelley and White poll. Daniel Yankelovich drew his inspiration from David Naisbett's book *Trend Report*, which was commissioned by the Club of Rome.

59. Michael Minnicino, Frankfurt School and 'Political Correctness', *Fidelio*, Winter 1992

60. Turn off your TV, Lonnie Wolfe, *New Federalist*, p.35, 1997

61. Ibid

62. Lincoln's American System Vs. British-Backed Slavery, Anton Chaitkin, *EIR*, February 15, 2008, p.42

63. The Classical War Against Multiculturalism: Brahms' Compositional Method, Dennis Speed, *Fidelio Magazine*, Winter 1993

64. http://en.wikipedia.org/wiki/The_Happening_%282008_film%29

65. Turn off your TV, Lonnie Wolfe, *New Federalist*, p.61, 1997

66. Ibid

67. Cleaning Up Talk Tv Is Tough When Fame, Shame And Sin Are So American, Maureen Dowd, *New York Times*, November 1, 1995

68. Ibid

69. INSNA: 'Handmaidens Of British Colonialism', David Christie, *EIR*, December 7, 2007, p.27

70. Theodore Adorno, *Television and the Patterns of Mass Culture*, 1954

71. Turn off your TV, Lonnie Wolfe, *New Federalist*, p.30, 1997

72. Turn off your TV, Lonnie Wolfe, *New Federalist*, p.22, 1997

73. Ibid

74. Turn off your TV, Lonnie Wolfe, *New Federalist*, p.25, 1997

75. Turn off your TV, Lonnie Wolfe, *New Federalist*, p.25-26, 1997

76. Ibid

77. Turn off your TV, Lonnie Wolfe, *New Federalist*, p.87-88, 1997

78. Ibid

79. Edward Bernays, *Propaganda*, 1928, reprint, Ig Publshing, 2004

80. http://en.wikipedia.org/wiki/Sesame_Street

81. http://en.wikipedia.org/wiki/Sesame_Street#cite_note-gladwell-100-24

82. http://en.wikipedia.org/wiki/Sesame_Street#cite_note-gladwell-91-26

83. Huston, Aletha C; Daniel R. Anderson, John C. Wright, Deborah Linebarger, & Kelly L. Schmidt (2001). ""Sesame Street Viewers as Adolescents: The Recontact Study." In Shalom M. Fisch & Rosemarie T. Truglio. "G" is for Growing: Thirty Years of Research on Children and Sesame Street. Mahweh, New Jersey: Lawrence Erlbaum Publishers. p. 133. ISBN 0-8058-3395-1.

84. http://en.wikipedia.org/wiki/Sesame_Street#cite_note-morrow-68-42

85. Ibid

86. Turn off your TV, Lonnie Wolfe, *New Federalist*, p.31, 1997

87. Turn off your TV, Lonnie Wolfe, New *New Federalist*, p.31, 1997

88. TALKING TELLY 3 – Portobello Film Festival. http://www.portobellofilmfestival.com/telly/telly003.html

89. Turn off your TV, Lonnie Wolfe, *New Federalist*, p.65, 1997

90. Don't hate me because I am beautiful, http://www.sociology.org/hate-beautiful/, August 16, 201

91. Harley Schlanger, Who Owns Your Culture, *Fidelio*, Spring 2003

CHAPTER SIX

Cybernetics

How does a utopia emerge, asks a voice-over in a German documentary simply titled, *The Net*. Does it come into being by chance, are there one or more inventors or is there a plan? It is at MIT, Massachusetts Institute of Technology, that an American and international science and engineering elite are educated. MIT also leads the way in the close partnership between the military and the university system. This collaboration began in World War I and continued throughout World War II when technology became the deciding factor in the war.

On August 13, 1940, the German Luftwaffe began the 'Battle of Britain.' Shortly after the start of the German bombings, Chicago-born mathematician and physicist, Norbert Wiener, (1894-1964), offered his expertise in the fight against fascism. Wiener, a 'pioneer' of information theory and coiner of the term *cybernetics*, was professor of mathematics at MIT "and had already dealt with questions of ballistics and artillery during World War I."[1] He was grappling with the problem of how to build a machine that could calculate in advance the movement of fighter planes, so that you could shoot them down.

To do that, "Wiener took into account the nature of technological warfare in which people, ships and planes are just abstract blips on the radar screen. The pilot becomes one with his plane, the boundary between man and machine becomes blurred, and what emerges is a mechanized, anonymous opponent whose actions can be modeled in war laboratory."[2] Although Wiener's machines were not operational until after the end of the war, the initial development of cybernetics was itself a by-product of war research.

While working on an anti-aircraft predictor, Wiener established parallels between the operation of servomechanisms, analogue control devices used in anti-aircraft gunnery and purposeful behaviour of pilots and gunners: in both cases, the goal was being reached via a feedback mechanism. From this research, Wiener then "postulated that control via feedback and communication via information exchange constituted universal mechanisms of purposeful behaviour for both living organisms and self-regulating machines such as computers."[3]

In fact, just as the Pentagon's Defense Advanced Research Projects Agency (DARPA)-funded experiments in the 1970s laid the basis for the Internet, innovations in science and technology such as Artificial Intelligence, computer graphics, holographics, satellite communications, fiber-optic cable TV, three-dimensional imaging, and the first generation of computer simulators created in places like MIT's famous Radiation Laboratory were by-products of the war effort[4] and served as critical think tanks for the U.S. military.

"The basis of cybernetics is the assumption that the human nervous system does not reproduce reality, but calculates it. Man, now appears to be no more than an information-processing system; thought is data processing and the brain is the machine made of flesh. The brain is no longer the place where 'ego' and 'identity' are mysteriously created through memory and consciousness. It is a machine consisting of switches and controlling circuits, feedback loops and communication nodes. A black box where cause is effect and effect is cause within an infinite cycle... a closed feedback system with input and output that can be controlled and calculated, no longer, as previously, starting out from the contemplation of nature, but from indisputable mathematics and logic."[5]

One gets a sense of being inside the Matrix trilogy staring at the cyberworld through the looking glass. Listen to *Matrix Reloaded*: "The new philosophy of human interrelations, sociometry, gives us a methodology and guide for determination of the central structure of society through the evocation of spontaneity of the human subject-agents. These factors, once located and diagrammed, supply us with the basis upon which the planning of all the many facets and activities of society may be undertaken – from juvenile and adult education to super-governments and world states."[6]

When compared to Wiener's own writing, the visions are remarkably similar. Now, listen to what Wiener had to say: "Very many of the factors which previously precluded a World State have been abrogated. It is even possible to maintain that modern communication, which forces us to adjudicate the international claims of different broadcasting systems and different aeroplane nets, has made the World State inevitable." Wiener was joined in this endeavour by many of the leading 'social scientists' and proponents of the counterculture movement such as anthropologist Margaret Mead and psychiatrist Gregory Bateson, who also participated in the Cybernetics Conferences of the 1940s, hosted by the Josiah Macy, Jr. Foundation.[7] Wiener, Bateson, Mead and company see cybernetics as the most revolutionary way of making science, perhaps the biggest bite out of the Newtonian "apple of knowledge" that mankind has taken in the past 2,000 years.

It was Wiener's sinister attack on the epistemology responsible for mankind's development and survival, which he elaborates in his book, *Cybernetics*, where he equates the human brain to a logical binary system,[8] that these social engineers found particularly useful.

In that, Wiener is Bertrand Russell's intellectual equal. Russell,[9] an embittered defender of oligarchic racialism, whose only love became the hatred of mankind had as his life-long mission "to reduce the human mind to a binary processor. This reductionism was the basis for experiments carried out by facilities such as the Tavistock Institute in London."[10] Extrapolating the premise into the counterculture, social engineers thought that computers could play a similar role to LSD for use in mind control. In other words, creating artificial 'concentration camps without tears.'"[11]

However, in order to have a 'concentration camp without tears' you must have control, communications and feedback. And that's where cybernetics come in with the behaviour of whole, complex systems known today as social engineering. It is a living history of social engineering that concerns us and how it evolved from old-fashioned electroshock therapy, to the modern day, "group shock therapy." MIT's Research Center for Group Dynamics and Britain's Tavistock Institute have been at the forefront of research into the subject of consensus through group dynamics. One thing empire builders like David Rockefeller have always known: Persuasion is key to setting up fascist movements. That's one of the reasons why initially, going back to the 1950s, oligarchical foundations were investing exorbitant sums of money in cybernetics and why the same foundations are spending billions today on social-networking technologies, in order "to change what people believe and what they do."[12]

But they weren't the only ones. Others carried out their research as well, such as Douglas McGregor, management professor at MIT and Alex Bavelas, a noted psychologist with Group Networks Laboratory at MIT, "who set up their own psychological laboratory, where they studied groups of managers through two-way glass. Forrester and McGregor wanted to literally map out the processes by which people change their views and opinions. The ultimate goal was to form a universal model of group decision-making. In 1964 Jay Forrester – a leading computer expert and the father of system dynamics who had joined the Sloan School in 1956 – even tried to design a model to create a computer simulation of a T-group sessions."[13] These programs could then be used to steer or "herd" popular opinion into a desired direction.

Over the years, the cybernetic "change agents" expanded their research from group think and decision-making models to developing technologies to map the flow of rumors through society, which they claim spread like the transmission of epidemics, such as AIDS.[14]

This technology, with minimal adjustments, could also be used to create social movements, thereby setting the stage for gang and counter-gang conflicts[15] – techniques entirely coherent with those used by Rockefeller-CIA and described in great detail elsewhere in this book.

The similarity to *Matrix Reloaded* is eerie and uncanny at the same time: "The task of the social scientist (change agents) is to invent the adequate tools for the exploration of a chosen domain. On the level of human interrelationships, this domain is made up of the interactive spontaneities of all the individuals composing it. Therefore, the task of the social scientist becomes the shaping of the tools in the fashion as to enable him to arouse the individual to the required point of spontaneity on a scale which runs all the way to the maximum. But individuals cannot be aroused – or only to an insignificant degree – by undynamic or automatic means. The individuals must be adequately motivated so that the full strength of their spontaneous responses is evoked. Thus, the intention and shaping of methods for social investigation and the stirring up of reactions, thoughts and feelings of the people on whom they are used must go hand in hand."[16]

The investigation and research into various social disciplines affecting human behavior and group think have always been funded by foundations with strong links to a control apparatus such as the Rockefeller Foundation, Ford Foundation, Russell Sage Foundation and Josiah Macy Foundation, to name a few.

In fact, "non-profit foundations have long been a favorite instrument by which one or another oligarchical faction has been able to discreetly test and implement new ideas in social control. The power of the biggest foundations is legendary. Since the end of the last century, institutions like the Ford Foundation, the Rockefeller Foundation, or the Russell Sage Foundation have been able to routinely override the objections of elected officials and go on to completely shape America's education policy, its public health policy, and even the operations of the Federal government itself."[17]

FIRST CYBERNETICS CONFERENCE. 8-9 MARCH, 1946, NYC

Cybernetics' origins can be associated with a series of 10 conferences throughout the 1940s and the 1950s, sponsored by the Josiah Macy, Jr. Foundation from 1946 through 1953. The foundation, created in New York City in 1930, was named after a New England Quaker named Josiah Macy, Jr. In 1872, the Macys' Long Island Oil Company became an arm's-length proprietary for the Rockefeller Foundation when they became part of the growing Standard Oil Empire of John D. Rockefeller.

Known among its members as the 'Man-Machine Project,' it was launched in May 1942 at a New York City conference called the Cerebral Inhibition Meeting, under the auspices of the Josiah Macy Foundation and sponsored by the medical director, Frank Fremont-Smith. In a sense, this was the foundation's first voyage into 'mind control.' The Cerebral Inhibition Meeting starred a leading expert on hypnosis named Milton Erickson. The discussion centerd around "problems of central inhibition of the nervous system that helped lay the basis for what would later become cybernetic theory."[18]

Among the participants were Norbert Wiener, Warren McCulloch, a professor of psychiatry and physiology at the University of Illinois and Kurt Lewin, of Frankfurt School. "Lewin's National Training Laboratory would later become part of the National Education Association, and would facilitate the transformation of public education in America into an approximation of Bertrand Russell's nightmarish scheme for teaching children that 'snow is black.'"[19] Others who attended the conference were Mexican physiologist Arturo Rosenblueth, behavioural scientist Lawrence K. Frank, Gregory Bateson, a pivotal player in MK-ULTRA and other secret government experiments with mind-altering drugs, including Margaret Mead, the assistant curator of ethnology at the American Museum of Natural History in New York, who would help launch the modern feminist movement. For all her goofiness, a sex-crazed Mead was heavily involved with the world of intelligence beginning with her work at Institute for Intercultural Studies, which had been financed during World War II by the Office of Naval Research as well as the Air Force's RAND Corporation.

According to Macy's web page, "It was Arturo Rosenblueth's presentation at an informal 1942 meeting of ideas he'd been developing with Norbert Wiener and Julian Bigelow that drew everyone's attention." Rosenblueth, a protégé of Norbert Wiener, outlined a conceptual agenda based on similarities between behaviours of both machines and organisms that were interpretable as being 'goal-directed.' This goal-directedness was framed in terms of definitive and deterministic 'teleological mechanisms.' 'Teleology' was transformed from philosophical mumbo-jumbo to concrete mechanism through the invocation of 'circular causality' in a system, whereby new behaviors were influenced by 'feedback' deriving from immediately preceding behaviors. This approach allowed one to address apparent purposiveness with reference to the present and the immediate past, without having to invoke references to possible or future events."[20]

According to Jeffrey Steinberg, "Rosenblueth proposed to draw together a group of engineers, biologists, neurologists, anthropologists, and psychologists, to devise experiments in social control, based on the quack claim that the human brain was nothing more than a complex input/output machine, and that human behavior could, in effect, be programmed, on both an individual and societal scale."[21]

However, these ideas would have to wait until after the end of WWII. Bateson undertook assignments in the pacific region with the US Army, while Rosenblueth and McCulloch returned to their research at MIT. The first conference tok place shortly after the war, in March 1946 in New York under the title, "The Feedback Mechanisms and Circular Causal Systems in Biology and the Social Sciences Meeting." All together, ten major conferences took place between 1946 and 1953.

Putting clinicians together with mathematicians, sociologists, and economists, the project's objective were "to create a theoretical model of extreme stress," especially taking into account the psychosomatic "feedback overload" which appeared to cause many shell shock cases.

From the cyberneticians' point of view, this was a truly revolutionary concept: "If one could create a model of a physiological system through which information is received from the environment, processed, and then fed back to change that environment, then, perhaps, it were possible to model the human mind itself, especially if one used the computational machines then being developed. The group became devoted to the premise, later stated by one of the founding members, John von Neumann, that the human nervous system is really just 'an efficiently organized, large natural automaton,' which is therefore subject to deterministic, linear mathematical modelling."[22]

"What came out of that first meeting was not only a demonic drive to create the ultimate engineered society, based on the fusion of man and machine. A core group of 20 people constituted themselves as a task force to carry out this mission, and would spawn a series of permanent institutions, where the work would continue, to the present day. A year after the founding session of the Macy project, Wiener would coin the term 'cybernetics' to describe their effort."[23]

In 1948, as a result of the first Macy Foundation meetings, Weiner wrote *Cybernetics: or the Control and Communication in the Animal and the Machine*.[24] Acclaimed to be one of the "seminal books... comparable in ultimate importance to... Galileo or Malthus or Rousseau or Mill," *Cybernetics* was judged by twenty-seven historians, economists, educators, and philosophers to be one of those books published during the "past four decades," which may have a substantial impact on public thought and action in the years ahead."[25]

One of the key discussion points at the very first Cybernetics conference dealt with the following question: Is the mind a product of the brain? From the animal world, Warren McCulloch and his sidekick Walter Pitts, two of the Macy meeting's organizers first tested networks of brain cells for their computation capabilities. From a cybernetic perspective, machines are not things but ways of behaving. To cyberneticians, such as Alan Turing and Joseph LeDoux, brains are 'universal machines' and 'devices for recording changes.' What all these people have in common is a fixed idea that our brains are a recursive set of cybernetic learning machines connected by a gigantic feedback loop to the universe itself. Consequently, the cybernetic, scientific paradigm embraces this type of relativity.

Ross Ashby, English psychiatrist and a pioneer in cybernetics, the study of complex systems explains: "Cybernetic, typically treats any given, particular machine by asking not what individual act will it produce now, but what are all the possible behaviors that it can produce. It deals with all forms of behaviour in so far as they are regular or determinate or reproducible. Natural reality is irrelevant. The connection between this approach to machines and the information theory is important. Information theory is characterised essentially by dealing always with a set of possibilities. What is important is the extent to which the system is subject to determining and controlling factors."

In other words, what cybernetics offer is the framework on which all individual machines may be ordered, related and understood. Thus, the larger the variety of actions available to a control system, the larger the variety of perturbations it is able to compensate. And how does Ashby define control? "As a reduction of variety: perturbations with high variety affect the system's internal state, which should be kept as close as possible to the goal state, and therefore exhibit a low variety."[26]

Weiner and Ashby argued that the sciences must be regarded as solely descriptive, devoid of cause. In this way, humanity's relationship to the world (and vice versa) becomes an erotic attachment to Aristotelian objects. The comprehension of the universal is therefore illusory. "The 'Unity of Sciences' attempted to destroy metaphysics and the existence of universal principles, by arguing that any divisions in science, e.g., any divisions between life, non-life, and cognition, were non-existent. They applied this extreme reductionism to physics and the social sciences alike, thereby claiming to unify them. Society was reduced to individual psychologies; individual psychologies were reduced to biological processes; biological processes were reduced to chemical processes. And so, human cognition was reduced to the electro-chemical processes of the brain: neurons firing or turning off, like a binary system. Finally, even the electro-chemical processes of the brain were reduced to Newtonian mechanics."[27]

By the early 1950s, Weiner, Ashby, British theorist Stafford Beer and others were grappling with the problem posed by Ashby: how to give aesthetics a firmly materialistic, non-causal basis. In essence, the cyberneticians' problem was Gottfried Wilhelm Leibniz. "At the beginning of the eighteenth century, Leibniz obliterated the centuries old Gnostic dualism dividing mind and body, by demonstrating that matter does not think. A creative act in art or science apprehends the truth of the physical universe, but it is not determined by that physical universe. By self-consciously concentrating the past in the present to effect the future, the creative act, properly defined, is as immortal as the soul which envisions the act."[28]

And this is something that the cyberniticians simply could not accept. They viewed cognition simply as a reaction to external stimuli. "Since bodies at rest stay at rest until acted upon by another body, the internal process of cognition was eliminated. Thus there was no 'divine spark', or soul. These conceptions would provide the basis for the discussions at the Cybernetics conference years later."[29]

Wiener and the cyberneticians thought the creative method was just a random by-product of access to 'information'. Therefore, "they would monitor the amount of information released into the 'field', acting as the information thermostat for society. In order to control the flow of information, they nested themselves inside major media outlets and opinion-shaping centers."[30]

This degenerate sleight-of-hand allows one to do several things at once. "By making creativity historically-specific, you rob it of both immortality and morality. One cannot hypothesize universal truth, or natural law, for truth is completely relative to historical development. By discarding the idea of truth and error, you also may throw out the "obsolete" concept of good and evil; you are, in the words of Friedrich Nietzsche, 'beyond good and evil.'"[31]

"Later, the heirs of the cyberneticians were involved in creating the 'information superhighway'. They created software that monitored the flow of "information" on the Internet like a massive electrical circuit board, setting up the circuit-breakers and monitoring the voltage. This concept was at the core of 'social networking', the establishment of sets of game theory matrices aimed at enforcing consensus. The mechanisation of societal relations was based on Wiener's idea that it were possible to mechanise thought."[32]

When applied to human models, we immediately run into problems. Humans are creative. Through the dominant, Judeo-Christian cultural matrix of Western ideology and the exercise of our reason, "we can discover principles beyond sense perception, and create technologies that allow our fellow humans to rise above the limits of our previous resource base. That is a simple refutation of the bogus entropy of Russell's positivists. We humans can also develop our mastery of social principles, like agape, in the domain of classical artistic composition. The ability to communicate these principles from one generation to the next enables a culture to elaborate its own continuing transformation. Modern nations can only achieve this progress by promoting the development of the sovereign minds of their citizens. Cultural development of this type, is the true mission of a republic."[33]

"The problem was, that as long as the individual had the belief – or even the hope of the belief – that his or her divine spark of reason could solve the problems facing society, then that society would never reach the state of hopelessness and alienation which was recognized as the necessary prerequisite for socialist revolution. The task of the Frankfurt School, then, was first, to undermine the Judeo-Christian legacy through an 'abolition of culture' and, second, to determine new cultural forms, which would increase the alienation of the population, thus creating a 'new barbarism.'"[34]

Therefore, Russell's intellectually degenerate offspring needed to eliminate those sovereign minds, otherwise their creativity would upset the equilibrium of the predetermined 'ecology'. "In Lewin's electromagnetic grid, those 'nodes' that attracted other 'nodes' through their ability to share ideas and create new capabilities for the survival of mankind, would need to be neutralized. This required the work of 'change agents', to bring the field back to the drab uniformity of consensus, and to maintain the equilibrium of ecology.

"Lewin's 'field theory', where 'field' simply became a statement that reflected assumed axioms about the nature of the relationships between the objects applied the same circular logic to human relations. And, as a closed system, the field was subject to the arbitrary laws of entropy. Lewin assumed that humans were like monkeys, whose relationships were determined through a calculus of hedonism. Cyberneticians such as Weiner, Ashby and Beer use a language of electromagnetics to describe the relationships, borrowing from James Maxwell's 'field theory' for electromagnetism. Since Maxwell had decided that causality in science was irrelevant, his theory wasn't actually a science.

"Where Maxwell assigned a "one" for a strong degree of cohesion and a "zero" for weak attraction in an electromagnetic grid, Lewin would do the same: "one" for the level of attraction between a monkey and its mother, "zero" for a predator monkey. The "field" became an aggregate of the relations among the hedonistic monkeys, which merely reflected Lewin's axioms about the nature of humanity. Universal principles, such as agape, were reduced to "game theory" by Lewin and his acolytes. As a closed system, devoid of principle, Lewin's field was also subject to entropy, or what a zoologist would call 'ecology.'

"Entropy applied to magnets and monkeys is one thing, but what happens when these rules are applied to humanity? Is a human economy subject to the same rules as monkey ecology?35 For Lewin, Maxwell, the Tavistockians, and all the intellectually sodomized children of Bertrand Russell, the answer is "yes!" It is here that our big problem arises, and it is here also, that these social engineers pulled off their masks to reveal their "fascism with a democratic face."36

For Norbert Wiener and his sidekicks, the core of the Cybernetics Group project, there was nothing sacred about the human mind, and the human brain was a machine, whose functioning could be replicated, and eventually surpassed, by computers.

FRANKFURT SCHOOL

One of the organizations directly linked to the cybernetics group was the Institute for Social Research (I.S.R.), but popularly known as the Frankfurt School. For example, Paul Lazarsfeld the director of the Radio Project was one of the invitees to the Macy Conferences. Also on hand was Max Horkheimer, the head of the Frankfurt School famous for his work in critical theory. "In May

1944, the American Jewish Committee established a Department of Scientific Research, which was headed by Frankfurt School director Max Horkheimer. Horkheimer established a project, called Studies in Prejudice, with generous funding from the AJC and other agencies, including the Rockefeller foundations.[37]

"The most significant of the five Studies in Prejudice, produced for the AJC during 1944-50, was *The Authoritarian Personality*.[38] Authors Adorno, Else Frenkel-Brunswik, Levinson, and Sanford assembled a large research team from the Berkeley Public Opinion Study and the International Institute of Social Research, to conduct thousands of interviews of Americans, to profile their allegedly deep-seated tendencies toward authoritarianism, prejudice, and anti-Semitism. Dr. William Morrow, the leading protégé of Dr. Kurt Lewin, who was one key, bridge figure between the Frankfurt School and the Tavistock Institute, was a research director for the Authoritarian Personality project."[39]

The authors of *The Authoritarian Personality* let it all hang out in the concluding chapter of the book, in which they summarized their findings and spelled out their recipe for social transformation:

"It seems obvious, that the modification of the potentially fascist structure cannot be achieved by psychological means alone. The task is comparable to that of eliminating neurosis, or delinquency, or nationalism from the world. These are products of the total organization of society and are to be changed only as that society is changed. It is not for the psychologist to say how such changes are to be brought about. The problem is one, which requires the efforts of all social scientists. All that we would insist upon is that in the councils or round tables where the problem is considered and action planned the psychologist should have a voice. We believe that the scientific understanding of society must include an understanding of what it does to people, and that it is possible to have social reforms, even broad and sweeping ones, which though desirable in their own right would not necessarily change the structure of the prejudiced personality. For the fascist potential to change, or even to be held in check, there must be an increase in people's capacity to see themselves and to be themselves. This cannot be achieved by the manipulation of people, however well grounded in modern psychology the devices of manipulation might be.... It is here that psychology may play its most important role. Techniques for overcoming resistance, developed mainly in the field of individual psychotherapy, can be improved and adapted for use with groups and even for use on a mass scale."

The authors conclude with this most revealing proposition: "We need not suppose that appeal to emotion belongs to those who strive in the direction of fascism, while democratic propaganda must limit itself to reason and restraint. If fear and destructiveness are the major emotional sources of fascism, *eros* belongs mainly to democracy."[40]

THE CYBERNETICS GROUP

The Frankfurt School and their closest allies at Tavistock, "were the architects of both the cybernetics project and the counterculture project of the 1960s. In fact, the Cybernetics Group, sponsored by the Josiah Macy Foundation, was the umbrella, under which the CIA and British intelligence conducted their mass experimentation with mind-altering psychedelic drugs, including LSD-25, which experiment was, eventually, spilled out onto the streets of San Francisco, New York's Greenwich Village, and every American college campus, giving us the counterculture 'paradigm shift' of 1966-72."[41] In 1965, one of the four directors of the Authoritarian Personality project, R. Nevitt Sanford, wrote the forward to *Utopiates: The Use and Users of LSD 25*, which was published by Tavistock Publications. I will come back to this later in the report.

If you are to believe the official gloss, since 1930, the Josiah Macy Jr. Foundation has devoted itself to the promotion of health and the relief of suffering, (http://www.josiahmacyfoundation.org/about/history),[42] directing its support toward medical education and established programs supporting conferences and publications on these and related topics.

The historian Jean-Pierre Dupuy describes the Macy Foundation's role in the Cybernetics Group this way: "Cybernetics was obliged from the beginning to ally itself with a movement- a political lobby, actually, operating under the auspices of the Macy Foundation – that sought to assure world peace and universal mental health by means of a bizarre cocktail concocted from psychoanalysis, cultural anthropology, advanced physics, and the new thinking associated with the Cybernetics Group."[43]

Corporate media propaganda subsequently brainwashed the masses to believe that Macy Foundation and figures such as Lazarsfeld and Weiner, Bateson, Mead and Ashby were merely participants in social experiments to improve the quality of life through the hodgepodge merging of technology, evolution and various social sciences. In reality, these individuals were lackeys for the foundations of the oligarchic World Empire – Rockefeller, Josiah Macy, Russell Sage and others.

Less well known to the general public is that the Macy Foundation has done more to harm America's social reality than any other single institution. "During and immediately after World War II, the Macy Foundation was utilized by a combination which included the British secret services, corrupt sections of American intelligence, plus treasonous members of the U.S. establishment, to conduct a vast social experiment. The overall purpose of the experiment was to end the uncontrolled optimism of a population that had just won a world war and had started to rebuild the world, and instead, to redirect that energy inward upon itself.

"Three parts of that experiment became so successful, that most people today think of them as some 'natural' evolution of Western society, rather than as an unnatural deformation imposed from the outside. The 'Sexual Revolution' of the late 1960s and early 1970s, which decisively accelerated today's collapse of the nuclear family, could not have occurred in the form in which it did, without the Macy Foundation's almost single-handed sponsorship of the development and dissemination of oral contraceptives through the work of Harvard's Dr. Gregory Pincus. Pincus was studying reproduction, and the eugenicists at Macy were studying how certain types of people should not be allowed to engage in reproduction. In 1954, Macy awarded Pincus a large special grant. In 1955, Pincus patented 'the Pill.'"[44]

"Similarly, the Macy Foundation was a primary actor in creating the 'Psychedelic Revolution,' thereby turning a post-war population that hitherto looked at sleeping pills with suspicion, into today's America, which routinely takes a new drug for every new mood, and prescribes psychoactive substances to its children by the millions of doses. And perhaps most importantly, it was the Macy Foundation that helped to pervert American citizens' world-renowned sense of technological optimism, into the now-pervasive ideology of the 'Information Age.'"[45]

THE MACY FOUNDATION AND MK-ULTRA

The bizarre Macy cocktail blend of the late 1940s and early 1950s became one of the primary covert funding conduits for the CIA through Macy Foundation director Frank Fremont-Smith. Fremont-Smith was very close to Dr. Harold Abramson, the CIA-linked psychiatrist and associated professionally with both Columbia University and Mt. Sinai Hospital whose LSD work had Macy Foundation cover.

One such individual was Dr. Louis Jolyon 'Jolly' West, then chair of UCLA's Department of Psychiatry and director of the Neuropsychiatric Institute. West was most famous as a CIA-contract psychiatrist whose work on the MK-ULTRA program in the 1950s involved feeding LSD to elephants.

Aside from Dr. Abramson, a significant number of the Macy Foundation conference members such as Gregory Bateson, Margaret Mead, and psychologist Kurt Lewin worked closely with the U.S. government during the Cold War under covert MK-ULTRA drug brainwashing experiments on the one hand and psychedelic drugs as a means of social manipulation on the other.

Another key participant in the Macy conferences was Donald Marquis, then a University of Michigan psychologist. "Marquis helped organize basic research on the psychology of propaganda, on the nature of various national targets, and on various social groups within nations through MIT's Center for Group Dynamics (CGD)."[46] It would inspire MIT's 'Group Networks Laboratory' to study small group dynamics under the supervision of Alex Bavelas, one of the top level advisor on Cold War secret operations for the Office of Naval Research (ONR) contracted out of MIT. This same MIT group later worked with the CIA to produce a major study of Communist Chinese "coercive persuasion" techniques more popularly known as "brainwashing."

Other Macy studies, such as a sponsored study in 'cell biology' were actually analysing techniques in eugenics, or 'race science.' Large amounts of grant money also went to what the foundation called 'psychosomatic interrelations,' that is, "how physiological change affects the mind, and vice versa. This was a cover for work on clinical techniques that would later be called 'brainwashing.'"[47]

"The Macy Foundation also provided financing and publicity for the British social engineer Dr. William Sargant, whose 1957 book,

Battle for the Mind, provided a "how-to-do-it" manual for mass brainwashing. Sargant, a specialist in 'shock trauma spent 20 years in the United States, working on the MK-ULTRA project and other secret mind-control efforts of the U.S. and British governments."[48]

EUGENICS

Throughout history, there have been those who, with an eye to specific political objectives have used terror or the threat of terror against targeted populations. The scientific rationale for tyranny has always been attractive to the elites because it creates a convenient excuse to treat their fellow men as lower than animals. Eugenics, a crackpot notion of hereditary superiority and inferiority originated in the 1880s and 1890s, spawned by a British network of families including Darwin's cousin Sir Francis Galton, Thomas Huxley, Sir Arthur Balfour, the Cadbury and the Wedgewood families as well as other late 19th century British Empire strategists linked to the Round Table movement of Cecil Rhodes and Lord Alfred Milner. They saw an opportunity to advance mankind towards a new Dark Age by taking the reigns of Darwin's perversely racialist 'survival of the fittest' natural selection evolution theory and applied social principles to develop social Darwinism.

In the United States, the story of eugenics begins in 1904, when the Cold Spring Harbor Laboratory was started by prominent eugenicist Charles Davenport with the funding of leading American would-be oligarchs, the Rockefeller, Carnegie and Harriman. By 1910, the British had created the first network of social workers expressly to serve as spies and enforcers of the eugenics race cult that was rapidly taking control of Western society. Hitler's English financial backers weren't the only ones sponsoring the eugenics research. In the 1920s, the Rockefeller family bankrolled the Kaiser Wilhelm Institute for Genealogy and Demography, which later would form a central pillar in the Third Reich.

At the end of the war, with corpses still smoking across Europe, the allies protected from prosecution the very Nazi scientists such as Josef Mengele who had tortured thousands of people to death. The Nazi radical brand of eugenics had embarrassed the Anglo-American social controllers, making 'eugenics' and 'mental hygiene' dirty words. The controllers, however, wouldn't be deterred. "In

1956, the British Eugenics Society decided in a resolution that, the society should pursue eugenics by less obvious means."[49] This means 'planned parenthood' and the environmentalist movement. Every population control policy was simply renamed as they continued their work under the protection of the United Nations and its associate organizations. The Eugenics, Euthanasia and Mental Hygiene Societies of Britain, America and the rest of Europe, was simply renamed the more palatable Mental Health Association of Great Britain and National Association of Mental Health of the United States which later became World Federation of Mental Health. I will come back to them later in this section.

Eugenics Quarterly Magazine became *Social Biology* and the American Birth Control League became Planned Parenthood, which today is responsible for a massive de-population in Africa. It is not widely known that some of the largest aid agencies and U.S. Christian Fundamentalist groups have been covertly running in Africa over the past several decades. Its banner is Family Planning, which is turned on its head when one understands the real implications and far-reaching objectives. These family planning policies are vigorously and consistently advocated by major bilateral donors such as the U.S. Government through its surrogate, U.S.-AID and multilateral agencies, most notably the International Planned Parenthood Federation (IPPF), the U.N. Fund for Population Activities (UNFPA) and the World Bank in Africa.

The World Bank, since the 1960s has been the leading funding entity of population control, with annual spending skyrocketing from a paltry $27 million in 1969-1970 to over $4.5 billion by 2006. World Bank Presidents Eugene Black and Robert McNamara were at one time trustees of Rockefeller-controlled Ford Foundation. "More significantly, population control policies have now become a key condition required for the disbursement of structural adjustment loans (SALs) by the World Bank and the International Monetary Fund within the framework of their Structural Adjustment Programs (SAPs)."[50] SAP conditions typically include, in addition to devaluation, liberalization and privatization of the national economies and of the health and education sectors, population control policies. Thus, preparation of a Population Policy Statement is a typical condition for the disbursement of a SAP.

Betsy Hartmann, Director of the Population and Development Program at Hampshire College, coined a new term for such policies: Malthusian Eco-Fascism (MEF). She notes that the international aid community currently concentrates its population control efforts in Sub Saharan Africa, and family planning becomes the number one priority: "The overriding imperative of these internationally generated programs is to reduce population growth as fast and 'cost-effectively' as possible." As she rightly observes, in much of Africa where AIDS threatens tragic human and demographic consequences, the present emphasis on population control and de-funding of health systems amount to indirect triage."[51]

Needless to say, Africa is not the only continent where these brutal policies are enacted. In 1972, in an effort to deal with the so-called 'population emergency' in India, the World Bank funded a $21 million project which "resulted in millions of involuntary sterilizations and thousands of deaths."[52]

The confluence of eugenicists, Nazis, environmentalists and One World Company enthusiasts also converge through a secretive 1001 Club, comprised of the most ancient and powerful families of Europe, who finance covert Malthusian operations of the WWF in Africa. One such operation run by WWF, for example, was called Operation Lock, ostensibly to save the black rhino in South Africa. "John Hanks, WWF's Africa director financed a team of 'retired' British SAS commandos to infiltrate and sabotage so-called poaching rings through an outside paramilitary force instigating black-on-black violence between the African National Congress and Inkatha Freedom Party by carrying out targeted acts of violence, like the June 18, 1992 Boipatong massacre."[53] The massacre caused the African National Congress to walk out of the initial formal negotiations to end apartheid, accusing the ruling National Party of complicity in the attacks.[54] The goal was "to trigger a bloody civil war and prevent the end of the apartheid system and the reintegration of South Africa into the world community."[55]

WWF and its direct action terrorist arm, Greenpeace, as well as like-minded groups are not just some lunatic fringe that can easily be ignored; they are the shock troops of the oligarchy in their fight against humany. "Malthusian" law similar to what was proposed for the UN 1994 Cairo Population Conference is a theory in demography regarding population growth developed during the industrial revolution on the basis of the writings of Thomas Malthus' famous 1798 *On Population*, which was nothing more than a plagiarized version of Venetian monk Giammaria Ortes' 1790 publication, *Riflessioni sulla popolazione delle nazioni*. According to his theory, population expands faster than food supplies. Ortes is the intellectual author upon which the genocidal 1994 U.N. Cairo Population Conference draft was based.

Most people may find all of this mind boggling at the very least, but the cultural heroes from the bygone era such as Margaret Mead, Huxley brothers, and the late Carl Jung, chief promoter of the "collective racial unconscious," not only attended conferences year after year with the Nazi doctor responsible for some of the most heinous crimes against the human race, but wholeheartedly supported every "theory" which brought on the Nazi holocaust against humanity. For example, Jung was an editor in the pre-war years of *German Journal of Psychotherapy* with Dr. M.H. Goering, Hermann Goering's cousin and a participant in a "T4" euthanasia project: the elimination of 400,000 mental patients in German institutions.

WORLD FEDERATION OF MENTAL HEALTH

World Federation of Mental Health is a perfect example of this confluence of Rockefeller-CIA-Tavistock apparatus. Rees, anthropologist Margaret Mead, behavioural scientist Lawrence K. Frank, Josiah Macy executive Fremont-Smith and German sociologist Max Horkheimer were all in Paris together, in the summer of 1948, to launch the WFMH. "Although he had died the previous year, Kurt Lewin had been involved in the preparations for launching the Federation, through his involvement, under Frank, in the National Committee for Mental Hygiene, and the London-centerd International Committee for Mental Hygiene, with a half-dozen Cybernetics Group members on its board. Both bodies oversaw a network of over 4,000 'psychiatric shock troops,' in Rees's words, who would be at the heart of a world-wide social-engineering apparatus, penetrated into every community.

"Margaret Mead and Lawrence K. Frank, two pillars of the Cybernetics Group, authored the founding statement of Rees's World Federation of Mental Health (both Mead and Freemont-Smith would later succeed Rees as president), which they titled,

"Manifesto of the First International. Mead and Freemont-Smith bluntly wrote: The goal of mental health has been enlarged from the concern for the development of healthy personalities to the larger tasks of creating a healthy society.... The concept of mental health is co-extensive with world order and world community."[56]

There are three dominant features to WFMH's Manifesto. Firstly, Rees stressed the need to apply World War I and World War II military psychiatric experience as a paradigm "for a system of community mental health clinics, through which counterinsurgency psychiatrists could forcibly administer"[57] 'therapy' to large population groups. Another key point in the Manifesto stressed the need of placing psychiatrists in key positions, which would allow them to play determinant roles in administrative and social programs of government and industry. As investigator L. Marcos wrote back in 1974, "one should not be astonished that the late George Orwell conceived his *1984* novel after a period of exposure to a group of Rees' followers."[58]

Launched by the Cybernetics Group in 1948, the World Federation of Mental Health (WFMH) was among the nastiest projects launched by the world's leading sociologists, psychiatrists, and anthropologists, the so-called "psychiatric shock troops," whose first president, Brig. Gen. John Rawlings Rees, was the director of the Tavistock Institute, Britain's premier psychological warfare center. Carl Jung was elected vice president. Another individual closely associated with WFMH was the Nazi doctor Werner Villinger who sat on the White House Conference on Children and Youth in the United States in the 1950s. For as much as Jung's supporters would like to deny it, there is no bending the facts: Jung was not confused about whom he was backing – Hitler and the Nazis.

Meanwhile, immediately after the war, Sir Julian Huxley changed the name of their program for enforced birth control, zero economic growth, and the technology of mass mind control, and continued to apply the principles which created Nazi Germany's mass murder against the "racially unfit." In 1946, Huxley announced that, "even though it is quite true that any radical eugenic policy will be for many years politically and psychologically impossible, it will be important for UNESCO to see that the eugenic problem is examined with the greatest care and that the public mind is informed of the issues at stake so that much that now is unthinkable may at least become thinkable."[59] In 1974, Henry Kissinger, Huxley's intellectually sodomized son said, "depopulation should be the highest priority of foreign policy towards the third world."

Huxley was a founder of the British Eugenics Society, first Director-General of UNESCO which pushed for population reduction, and what Huxley called 'a single culture for the world'; he was also a leading member of WFMH, director of the Abortion Law Reform Association and founder of the World Wildlife Fund whose first president was the former card-carrying Nazi, Prince Bernhard of the Netherlands and one of the organizers of the Bilderberg meetings since its inception in 1954.

Few realize that Huxley represented a social set of miscreants the likes of which have not been seen since. Amongst them was his grandfather, Thomas Huxley, whose outspoken support of Charles Darwin's degenerate theory of evolution earned him the nickname, Darwin's Bulldog; his personal mentor and Satanist Aleister Crowley; the aforementioned Brigadier John Rawlings Rees; Bertrand Russell and H.G. Wells, whose *Open Conspiracy* was a call for fascist imperial world dictatorship. "What makes the 'open conspiracy' open, is not the laying out of some secret masterplan, not the revealing of the membership roster of some inner sanctum of the rich and powerful, but rather, the understanding that ideas, philosophy and culture, control history." Also involved, Julian's brother Aldous, of the *Brave New World* fame. In Aldous Huxley's book, the population is genetically divided into Alphas, Betas, all the way down to the synthetically produced Epsilon morons through the application of eugenics, or Nazi breeding laws.

Over the years, "the WFMH itself became a huge clearing-house of psychological warfare profiles used by British and American intelligence services during the Cold War. A typical Macy-WFMH joint project was the 'Conferences on Problems of Health and Human Relations in Germany,' a series of high-level, 1950-51 meetings designed to convince German social scientists and health care providers that the Frankfurt School's bogus 'authoritarian personality' profile should be relied upon in dealing with German patients."[60]

New terms like transhumanism, population control, sustainability, conservation, bioethics and environmentalism replaced the now outdated racial hygiene and social Darwinism and in the process "prepared public opinion, medical education, and government policy to discard the notion of the sanctity of human life."[61]

Except that as investigative journalist Rob Ainsworth explains, "the environmentalist movement is anything but the grassroots movement it pretends to be. The biggest and most influential groups receive tens of millions of dollars in funding every year, and the boards of trustees and directors reflect this."[62] From Survival International to United Nations Environment Program, from Worldwatch Institute to International Food Policy Research Institute, from World Resources Institute to Greepeace, the Nature Conservancy and Swiss-based International Union for the Conservation of Nature, from the UN Development program to the Natural Resources Defense Council, and Environmental Defense... are run by bankers, hedge-fund managers and big oil. As Club of Rome wrote in its 1991 publication, *The First Global Revolution*, "The real enemy is humanity itself."

TRANSHUMANIST AGENDA

Another area of globalist agenda is transhumanism. As I explain in *TransEvolution – The Coming Age of Human Deconstruction*, transhumanism is an ultra-high tech dream of computer scientists, philosophers, neural scientists and many others. It seeks to use radical advances in technology to augment the human body, mind and ultimately the entire human experience. It is the philosophy that supports the idea that mankind should pro-actively enhance itself and steer the course of its own evolution.

Transhumanists wish to become what they call "post-human." A post-human is someone who has been modified with performance-enhancing body and brain augmentations to the point that they could no longer call themselves human. They have mutated themselves into an all-together new being.

Many people have trouble understanding what the true transhumanism movement is all about, and why it's so awful. After all, it's just about improving our quality of life, right? Or is transhumanism about social control on a gigantic scale?

In fact, "social control" is exactly how powerful American foundations such as the Rockefeller Foundation, Carnegie

Endowment and Macy Foundation see it. The ability to make machines act like humans, and the ability to treat humans as machines: "...the final accomplishment of H.G. Wells' old Fabian goal of a 'scientific world order' where everything is as neat as a differential equation, and unpredictable things such as human creativity never mess things up."[63]

To most people, this sounds like something from a science fiction film. Few are aware of constant breakthroughs in technology, which makes the transhumanist vision a very real possibility for the near future.

For example, neurochip interfaces, computer chips that connect directly to the brain are being developed right now. The ultimate goal of a brain chip would be to increase intelligence thousands of time over – basically turning the human brain into the super computer.

Lifelong emotional well-being is also a key concept within transhumanism. This can be achieved through a recalibration of the pleasure centers in the brain. Pharmaceutical mood renders have been suggested, which will be cleaner and safer than the mind-altering drugs.

The goal is to replace all aversive experience with pleasure beyond the doubts of normal, human experience. Nanotechnology, for example, is a pivotal area concerning transhumanists. It is the science of creating machines, the size of molecules. Such machines could create organic tissue for medical use.

Using this type of technology could dramatically prolong life span. The Global Future 2045 International Congress predicted that by 2045, it will be possible to live forever by combining technology and biology into an event called: "The Singularity."

The Singularity would occur at the point in which artificial intelligence surpasses the capabilities of the human brain. From cyborgs with very long life spans, to downloading consciousness itself into a machine, transhumanists say that it is impossible to predict exactly what a post-human will be, *but that it will be better*. Such lofty promises are being embraced by many people – for a better world for everyone.

No matter how you look at it, Singularity is being promoted as *the* great solution to our 21st century global problems. And thanks to the human genome project, we will soon be able to decode DNA itself. Through the use of applied genetics, science will then be able to improve the human race. What most people don't realize is that this concept is not new.

Transhumanism was born out of humanism, which is yet another clever disguise of "scientism," created specifically so that global eugenics operations could be carried out without being noticed.

Some experts say that soon it might be possible to live forever. Artificial intelligence, the creation of thinking robots is closely related to the mind machine merger concept of neurochips. There is a growing perception that machine parts will be added to human bodies to create cyborgs, a term invented by Dr. Nathan Kline, a psychiatrist linked to the CIA and MK-ULTRA. He had co- written an article with Manfred Clynes for the September 1960 issue of *Astronautics* entitled "Cyborgs and Space" which first introduced the term 'cyborg' into the English language.

Still others believe that humanity will merge totally with technology by uploading individual consciousness to a virtual reality. Upon being uploaded, one could live forever within a computer-generated reality, leaving the physical body behind. In this machine, the individual could merge his intelligence with the collective intelligence of all others in the digital reality, effectively becoming one super intelligent being. This concept is popularly referred to as the 'hive mind.' How many people see the parallel with Jung's 'collective unconscious'? "Only through the recognition of the dimension of the collective unconscious can science serve the interests of man," says Jung. Except that virtual reality and computers cannot replicate human intelligent thought. What AI can do, is to alter consciousness, in the same way that LSD altered and degraded the human mind back in the 1960s.

In the concluding chapters of *Cybernetics*, Wiener states the possibility of a future with learning-capable, self-reproducing machines, much like that depicted by an Arnold Schwarzenegger cyborg, in the apocalyptic movie *Terminator*. But, like all computers or logical systems, all the decisions and policy of those machines will be nothing more than a logical deduction "from a set of rules and axioms of its initial programmer. There is no possibility for discovery of a new universal principle of science"[64] because the actual principles of human knowledge are not confined to what is learned.

As Johannes Kepler exhaustively explained in his *The Harmonies of the World*, "the human personality has access, by the unique nature of human mind's unique power of cognition," unique among all living creatures, "to generate previously unknown, validatable universal physical and comparable principles" which increase our species' power in and over the universe per capita and per square kilometre of space.

To annul this unique nature of human mind, the parallel world of Jungian thought can now be found intertwined with the ideas of New Age. From beer commercials that talk of the collective unconscious to the *Star Wars* series and mass marketed virtual reality, we are consciously building on Jungian imagery and mystical ideas. To them, " this new technology is the key to open the doors of perception to Jungian dream world."[65] However this Jungian *Star Wars* "dream world" is more like a recurring nightmare. The most crucial epistemological issue is often overlooked. Do these *Star Wars*, *Doom* and *Lord of the Rings* creatures look human to you? The second issue is how do you corrupt people into Jungian mysticism? Obviously, by dehumanizing the image of man. Therein lies the principle of evil pervading "virtual consciousness." Am I exaggerating? Let's see.

"If you can generate enough stimuli outside one's sense organs to indicate the existence of a particular alternate world, then a person's nervous system will kick into gear and treat the stimulated world as real," says Jaron Lanier of VPL Research, Inc. Furthermore, if virtual reality technology is administered as a brainwashing tool, it can weaken what the Freudians call the 'super-ego,' that part of the personality that coheres to one's moral conscience, which is the basis for truth seeking on the one hand and the uniquely human act of fundamental discovery and integration of new universal principles which have improved the lives of people per square kilometre of space against nature on the other. In fact, science has tried to drum out our built-in connection to the universe and replace it with 'consensual hallucination' through virtual reality.

Virtual reality, we are told, creates an artificial world in which we are free to do whatever we want without being accountable for our actions. If you want to have sex with the computer-generated image of your neighbor's dog, that's perfectly all right, go ahead and have it in the virtual world. With the liberation of our repressed fantasies, if one is to believe Freudian claptrap, virtual reality produces a new consciousness that redefines good and evil. In the words of that other famous psychopath, Friderich Nietzsche, it is a universe "beyond good and evil." Thus we are liberated to do evil "without suffering the effect of evil and therefore purge ourselves of our innate desire to 'be' evil."[66]

The fantasized pinnacle of transhumanist world provides us with the clearest view of its deadly ends, when seen through the eye of a physical world lacking universal truth and "the sacred responsibility of the sovereign individual to act for the Good."[67] And without universal truth, there can be no reason. If truth is killed, then our civilization is killed with it.

TRANSHUMANIST AGENDA & THE ENTERTAINMENT INDUSTRY

Another key issue of transhumanism is the fostering of a particular agenda as the basis of forms of entertainment. A simple example of this can be found in numerous Hollywood films or television shows, which include robots or androids. They often make the viewer feel sympathetic towards the robot and in turn this makes them feel emotional, if say, the robot were to be 'killed' by a human. Its also worth noting that it is almost always a human being causing these robots "pain."

For example, in *Terminator* and *Short Circuit*, this agenda is openly flaunted. The robotic characters in these films are shown to have good morals and human-like emotions throughout the entire movie. As a result, the viewer then, becomes attached and shows compassion and sympathy for them.

So, how does it relate to the music industry?

The universe and its composition are essentially made up of matter vibrating at different frequencies. In essence, the universe is made up of sound, and this includes us human beings, too. Sound is widely used as a therapeutic tool and can alter mental and physical states of human beings.

Music activates nearly every region of the brain. The limbic system in the brain begins to fire when we have a variety of pleasurable experiences. That same network of neurons also fires when we listen to music that we enjoy. These neurons help to modulate levels of the neurotransmitter dopamine in the brain. Dopamine is one of the neurotransmitters that eventually make you feel positive. So when you listen to music that you like, you get an actual change of chemical levels in your brain. And this is why the music industry is also being used to push the transhumanist agenda. By having nearly every region of the brain active, you become easily susceptible to psychological attacks, in this case, by your own favorite artists.

There are numerous examples of transhumanist agenda being pushed in the music industry. Take Rihanna's live performance of 'Rude Boy', for example. On stage, she performs with real dancing robots. These robots have absolutely nothing to do with the song. The purpose of that dance routine is to make the viewers feel as if the robots are sentient beings. The whole idea of the transhumanist agenda is to make humans view inanimate objects and technology as being on the same level as a human being.

Setting for a book: "God of Singularity springs out of a computer to rapture the human race. The math-magicians of British Intelligence calculate demons back into the dark. And solar-scale computation just uploads us all into the happy ever after."[68] The "God" is really just an Artificial Intelligence, just one of the plethora of post-humanizing changes and memetic subcultures that are coming on fast and that most people don't even know exist.

A cover of a well-known fashion magazine has a slimly clad young lady embracing a robot while meaningfully looking into his eyes. That's what sexualizing technology is all about. This agenda is not about the improvement of the human species but rather about the manipulation of the human consciousness, so that human beings no longer see themselves as human beings, but rather much like a piece of technology. By dehumanizing our species, we become the enemy. You may not realize it, but Tavistock has long used Hollywood and the music industry as two of its most potent tools in reaching out to the brainwashed masses in order to change the historical paradigm of society. The next time you look at an all-knowing grin of a Tavistockian degenerate, just remember, that's what they are thinking about.

Endnotes

1. Useless Eater Blog: The Net: The Unabomber, LSD And Internet, http://uselesseaterblog. blogspot.com/2012/04/net-unabomber-lsd-and-internet-stud

2. Ibid

3. Larouche Planet | Library / Chapter 4 Lyndon In Wiener World, http://laroucheplanet. info/pmwiki/pmwiki.php?n=Library.WienerWorld

4. MIT's then-President James Killian later became Chairman of the Board of Consultants on Foreign Intelligence Activities in 1955 as well as Chairman of the Science Advisory Panel for the U.S. Army

5. Useless Eater Blog: The Net: The Unabomber, Lsd And Internet, http://uselesseaterblog. blogspot.com/2012/04/net-unabomber-lsd-and-internet-stud

6. *Matrix Reloaded*, Warner Brothers, May 2003, Producer Joel Silver

7. See David Christie, "INSNA: 'Handmaidens of British Colonialism,' LaRouche PAC

8. Weiner, Cybernetics, Computing Machines and the Nervous System, ch. 5, *The MIT Press*, 1965.

9. Lyndon H. LaRouche, Jr., "How Bertrand Russell Became an Evil Man," *Fidelio*, Fall 1994.

10. David Christie, INSNA: 'Handmaidens of British Colonialism,' *EIR*, December 7, 2007

11. Creighton Cody Jones, Only Diseased minds believe in Entropy, How Weiner attemped to kill science, *EIR*, Jan 4, 2008

12. "The Stanford Persuasive Technology Lab creates insight into how computing products – from websites to mobile phone software – can be designed to change what people believe and what they do. For that reason, we're studying Facebook – it's highly persuasive." http://credibilityserver.stanford.edu/captology/facebook

13. Andrea Gabor, *The Capitalist Philosophers: The Geniuses of Modern Business – Their Lives, Times, and Ideas* (New York: Times Business, 2000), 163

14. Center for Models of Life, out of Niels Bohr Institute. http://cmol.nbi.dk/models/inforew/inforew.html

15. David Christie, INSNA: 'Handmaidens of British Colonialism,' *EIR*, December 7, 2007

16. *Matrix Reloaded*, Warner Brothers, May 2003, Producer Joel Silver

17. Michael Minnicino, Drugs, Sex, cybernetics, and the Josiah Macy Jr. Foundation *EIR*, July 2, 1999, Vol 26, number 27

18. Larouche Planet | Library / Chapter 4 Lyndon In Wiener World, http://laroucheplanet. info/pmwiki/pmwiki.php?n=Library.WienerWorld

19. Jeffrey Steinberg, "From Cybernetics to Littleton: Techniques of Mind Control," *EIR*, May 5, 2000.

20. http://www.asc-cybernetics.org/foundations/history2.htm

21. Lyndon H. LaRouche, Jr., Star Wars and Littleton, *EIR*, June 11, 1999

22. Drugs, Sex, Cybernetics and the Josiah May, Jr. Foundation, Michael Minnicino, *EIR*, Volume 26, Number 27, July 2, 1999, p.37

23. Jeffrey Steinberg, "From Cybernetics to Littleton: Techniques of Mind Control," *EIR*, May 5, 2000.

24. Norbet Wiener, *Cybernetics or Control and Communication in the Animal and the Machine* (New York: John Wiley, 1948), 9-13.

25. John B. Thurston, *The Saturday Review of Literature*

26. http://pespmc1.vub.ac.be/REQVAR.html

27. David Christie, INSNA: 'Handmaidens of British Colonialism,' *EIR*, December 7, 2007, p.32

28. Michael J. Minnicino, *The New Dark Age: The Frankfurt School and Political Correctness*, Winter 1992

29. David Christie, INSNA: 'Handmaidens of British Colonialism,' *EIR*, December 7, 2007, p.32

30. Ibid

31. Michael J. Minnicino, *The New Dark Age: The Frankfurt School and Political Correctness,* Winter 1992

32. David Christie, INSNA: 'Handmaidens of British Colonialism,' *EIR*, December 7, 2007, p.32

33. Ibid

34. Michael J. Minnicino, *The New Dark Age: The Frankfurt School and Political Correctness,* Winter 1992

35. David Christie, INSNA: 'Handmaidens of British Colonialism,' *EIR*, December 7, 2007

36. See, for instance, the November-December 1 974 issue of *The Campaigner*, "Rockefeller's 'Fascism with a Democratic Face,' " ICLC Strategic Study

37. Jeffrey Steinberg, "From Cybernetics to Littleton: Techniques of Mind Control," *EIR*, May 5, 2000

38. Theodor W. Adorno, Else Frenkel-Brunswik, Daniel J. Levinson, *The Authoritarian Personality: Study in Prejudice*, New York: Harper, 1950. Reprinted by W. W. Norton & Co Inc; Abr Rei edition, 1993

39. Jeffrey Steinberg, "From Cybernetics to Littleton: Techniques of Mind Control," *EIR*, May 5, 2000

40. Ibid

41. Ibid

42. http://www.asc-cybernetics.org/foundations/history2.htm

43. Jean-Pierre Dupuy, *Mechanization of the Mind*, p.23

44. Michael Minnicino, Drugs, Sex, cybernetics, and the Josiah Macy Jr. Foundation *EIR*, July 2, 1999, Vol 26, number 27

45. Ibid

46. Larouche Planet | Library / Chapter 4 Lyndon In Wiener World, http://laroucheplanet. info/pmwiki/pmwiki.php?n=Library.WienerWorld

47. Michael Minnicino, Drugs, Sex, cybernetics, and the Josiah Macy Jr. Foundation *EIR*, July 2, 1999, Vol 26, number 27

48. Jeffrey Steinberg, "From Cybernetics to Littleton: Techniques of Mind Control," *EIR*, May 5, 2000

49. The Genocidal Lombard League of Cities Apparatus, *EIR Dossier,* June 6, 2008, pp.50-55

50. Martin, Guy, The West, Natural Resources and Population Control Policies in Africa Historical Perspective, *Journal of Third World Studies*. Spring 2005, pp. 16-18

51. Hartmann, Betsy, Cross Dressing Malthus, *ZNet Commnetary*, Sept 23, 1999

52. Elizabeth Liagin, Excessive Force: power, politics and population control, *Information Project for Africa*, 1996

53. The Coming Fall of the House of Windsor, *The New Federalist*, 1994

54. Boipatong Massacre. ANC 1992-06-18. http://www.anc.org.za/ancdocs/pr/1992/pr0618. html. Retrieved 2007-04-28.

55. *Wars in Africa: the final stage of globalization*, Uwe Friesecke, February 26, 1999, p.51

56. Jeffrey Steinberg, "From Cybernetics to Littleton: Techniques of Mind Control," *EIR*, May

5, 2000

57. Michael Minnicino, Low Intensity operations: The Reesian Theory of war, *The Campaigner*, April 1974

58. Ibid

59. UNESCO Chair Admits Organization Was Founded To Push Global Governance, Jurriaan Maessen, InfoWars.com, August 21, 2010, http://www.infowars.com/unesco-chair-admits-organization-was-founded-to-push-global-governance/

60. Michael Minnicino, Drugs, Sex, cybernetics, and the Josiah Macy Jr. Foundation *EIR*, July 2, 1999, Vol 26, number 27

61. The Nazi Euthanasia Program: Forerunner of Obama's Death Council, Anton Chaitkin, http://archive.larouchepac.com/node/10654, June 15, 2009

62. Rob Ainsworth, The New Environmentalist Eugenics: Al Gore's Green Genocide, March 30, 2007, *EIR*

63. Michael Minnicino, "Drugs, Sex, cybernetics, and the Josiah Macy Jr. Foundation," EIR, July 2, 1999, Vol 26, number 27.

64. Only Diseased Minds Believe in Entropy, Creighton Cody Jones, *EIR*; January 4, 2008

65. Lonnie Wolf, Turn off your TV, *The New Federalist*, p.87

66. Lonnie Wolf, Turn off your TV, *The New Federalist*, p. 84

67. Ibid

68. RU Sirius & Paul McEnery, The Reluctant Transhumanist, *h+Magazine*, April 1, 2009

CHAPTER SEVEN

Science Fiction
and the Tavistock Institute

Many literary scholars name Mary Shelley's *Frankenstein* as the first science fiction novel. Some of the early works with science fiction themes such as Nathaniel Hawthorne short stories shared a cautionary message that men should not use science to interfere with nature. Later, these ideas were transformed into science fiction films professing the idea of technology extinguishing humanity through Armageddon-like disasters such as Hollywood's *The Day The Earth Stood Still* (1951), *When Worlds Collide* (1951), the three blockbusters *Deep Impact* (1998), *Armageddon* (1998) and *The Book of Eli* (2010). They provide a grim outlook, portraying a dystopic view of the world in the advanced phase of decay such as *Metropolis*, filmed in 1927 with its underground slave population and view of the effects of industrialization on society.

Frankenstein may have been the first science fiction novel; however, the originator of the science fiction genre without a doubt was H.G. Wells. At the end of the 19th century H.G. Wells gave us the *War of the Worlds*, which in 1953, over half a century later, would be transformed into a Hollywood blockbuster. Other notable science fiction titles by Wells are *The Time Machine*, *The Island of Dr. Moreau* (1886) and *The Invisible Man*. But it took the television franchise, *Star Trek* to bring science fiction into the mainstream. *Star Trek* made travel through space more plausible than ever.

However, a closer look at the themes of science fiction demonstrates that the entire genre is, not quite the source of collective enlightenment and a body of imaginative literature it is made out to be, but rather, oriented towards something far more sinister in need of deconstruction. From stories of distant planets, unknowable multi-dimensional universe, inexplicable forces, strange and extraordinary microscopic organisms or mutant monsters unleashed on the human race, either created by misguided mad scientists or by nuclear havoc such as in *The Beast From 20,000 Fathoms* (1953), impossible quests, improbable settings, of "parallel worlds," of futuristic technology and alien intelligence far superior to man, not to mention disaster stories about technologies that backfire, form part of a concerted psychological assault on people's belief systems that the laws of the universe are rational and therefore knowable to the human mind.

These stories often portray the dangerous and sinister nature of knowledge ('there are some things Man is not meant to know') as is repeated continuously in countless 'B' movies such as *Frankenstein*, (1931), *The Island of Lost Souls* (1933), including the threatening, existential loss of personal individuality such as *Invasion of the Body Snatchers* (1956), nuclear annihilation in a post-apocalyptic world in *On the Beach* (1959), are all assertions of the impotence of the human mind before the vast, unknowable, and uncontrollable cosmos. Plots of space-related conspiracies such as *Capricorn One* (1978), supercomputers threatening impregnation in *Demon Seed* (1977), the results of germ-warfare in *The Omega Man* (1971) and laboratory-bred viruses or plagues such as *28 Days Later* 2002); then we have black-hole exploration as in *Event Horizon* (1997), and futuristic genetic engineering, human transformation and cloning in *Gattaca* (1997) and Michael Bay's *The Island* (2005).

As science fiction writers themselves put it, their stories are designed to 'bend the minds' – in other words, to destroy them. And, no wonder. Commonly, sci-fi genre expresses society's anxiety about out-of-control technology, self-augmenting humanoid robots millions of times smarter than mankind, aliens determined to destroy humanity, kill switches and Artificial Intelligence protocols, depicting an eternal struggle of good versus evil that is played out by recognizable archetypes and warriors such as in *Forbidden Planet* (1956) and the space opera *Star Wars* (1977) with knights and a princess with her galaxy's kingdom to save.

If we are to believe Hugo Gernsback, founder of *Amazing Stories*, and the first specialist science fiction magazine in America said that "Not only is science fiction an idea of tremendous import, but it is to be an important factor in making the world a better place to live in, through educating the public to the possibilities of science on life which, even today, are not appreciated by the man on the street.... If every man, woman, boy and girl, could be induced to read science fiction right along, there would certainly be a great resulting benefit to the community, in that the educational standards of its people would be raised tremendously. Science fiction would make people happier, give them a broader understanding of the world, make them more tolerant."[1]

However, science fiction is not a literary form. It originated as, and remains to this day, a sophisticated tool designed to entrap, disorient, and destroy the promising creative minds of the younger generation who have 'dangerous' illusions that are associated with belief in the idea of technological progress and the spirit of innovation. Once the belief in the power of human reason is gone, all potential for future scientific work is destroyed. The current explosion in the visual media has played a decisive role in modern science fiction/fantasy films such as *Lord of the Rings* and *Avatar*. Today, more than half of Americans believe in the existence of UFOs and an entire movement has been built on the basis of replacing science and technology with futuristic scenarios.

A recent CNN/*Time* Poll provides what is possibly the most dramatic confirmation of the destructive effects of almost a century worth of promotion of science fiction on the population. According to the poll, 80% of Americans believe that the government is hiding the existence of extraterrestrial life forms. 64% believe aliens have contacted humans. 50% believe that aliens have abducted humans. 75% believe that a UFO crashed near Roswell. 26% believe that we should expect to be treated as an enemy. 39% believe that aliens would appear very humanoid. 35% believe that aliens would look somewhat human. Almost 22% believe they have seen an Unidentified Flying Object themselves. Amongst the last group of people, we have former U.S. President Jimmy Carter who allegedly saw one such object over his peanut farm in Georgia back in his pre-presidential days.

Few people realize that the genre was laboratory-designed and imposed on the minds of American youth by the same financial and political interests who earlier sponsored the drug trade and later, the counterculture movement in the 1960s. What is perhaps even more surprising, is that special interest groups with far-sighted agendas financed the world's most recognizable names.

THE EARLY DAYS

H.G. Wells, head of British foreign intelligence during World War I and the spiritual grandfather of the Aquarian Conspiracy, was a protege of leading Darwinist T. H. Huxley, founder of the British Round Table intelligence organization along with Cecil Rhodes. The Round Table is one of the 'open secret' secret societies directly linked to the Black Guelphs, better known as the House of Windsor. Another, is the Society of Carpocrates with its direct links to Queen Elizabeth II, a Black Guelph through her grandmother Queen Victoria. The most important member of British Round Table was Baron Harold Anthony Caccia whose family is one of the oldest families in the Black Nobility. In turn, Venetian Black Nobility was interlocked with a secret organization OC (Organizational Consul), controlled by the Thule Society, which functioned as the 'mother organization' of a plethora of parties, societies, paramilitary units and terrorist groups. The 'greatest' creations of the Thule Society were the Nazi Party and its degenerate leader, Adolf Hitler.

Wells, the novelist, was a member of the elite British oligarchic planning group, the Coefficients. Both the Round Table and the Coefficients were committed to the establishment of a "feudal empire run by an aristocracy which controlled all knowledge and technology and used them to rule over a population of ignorant, drugged plantation slaves."[2]

The Coefficients were "a cross between a diners club and a modern think-tank, which met monthly over dinners at London's St. Ermin's Hotel from 1902 to 1908. Among the members of this group was the powerful Lord Robert Cecil, elder statesman of Britain's most powerful family, and cousin to Arthur Balfour, then serving as Conservative Prime Minister. Lord Alfred Milner, the High Commissioner of South Africa"[3] also participated on a regular basis as well as Halford Mackinder, the newly appointed head of London School of Economics. Others involved were Earl Bertrand Russell as well as Sidney and Beatrcie Webb, Fabian socialists and supporters of Mussolini.

In other words, the crowd behind H.G. Wells, were some of the most powerful empire builders belonging to the upper echelons of secret societies and financial interests.

The Bilderberg Group, in fact was a natural extrapolation of the Coefficients Club. Milner spoke of his vision for the future during a 1903 meeting, over half a century before the Bilderberg Group was founded. At the meeting, Milner stressed the point:

"We must have an aristocracy – not of privilege, but of understanding and purpose – or mankind will fail.... And here my peculiar difficulty with democracy comes in. If humanity at large is capable of that high education and those creative freedoms our hope demands, much more must its better and more vigorous types be so capable.... The solution does not lie in direct confrontation. We can defeat democracy because we understand the workings of the human mind, the mental hinterlands hidden behind the persona.

"We need constructive imagination working within the vast complex of powerful people, clever people, enterprising people, influential people, amidst whom power is diffused today, to produce the self-conscious highly selective, open-minded, devoted aristocratic culture, which seems to me to be the next necessary phase in the development of human affairs. I see human progress, not as the spontaneous product of crowds or raw minds swayed by elementary needs, but as a natural but elaborate result of intricate human interdependencies, of human energy and curiosity liberated and acting at leisure, of human passions and motives modified and redirected by literature and art."[4]

In other words, educated masses are a horrible idea, because they would signify the death of oligarchism. Nations that foster the creative-mental developments of their population produce a people which will not tolerate oligarchical forms of rule indefinitely. Illiterate, technologically backward populations will. In fact, one of the reasons for the creation of the science fiction genre had to do with population density. By the middle of the 18th century, the Venetian oligarchy special interest groups throughout Europe realized that there was a direct relationship between levels of technological progress and population density.

As writer Ian D. Colvin said in his book, *The Unseen Hand in English History*, "The ruling motive in politics, as intelligent men know very well, is interest. Those resounding cries and plausible principles on which to the young and the innocent the battle appears to be fought, are usually pretexts, the flag, the colour of action; men and especially politicians, seldom reveal their true motive, but nearly always wrap it in some virtue, faith or plausible abstraction."[5]

Although anyone, who understands the deeper implications of the proverbial "corridors of power," knows that in politics hardly anything happens without unwritten deals over purposeful plans. Yet, often when one speaks of a 'stringpuller behind the scenes,' great howls of protest arise.

Absolutely inimical to the implementation of the Round Table's feudal one-world order was the development of sovereign republican nation-states, committed to the development of their industrial and scientific prowess and to the moral capabilities of their citizenry. "Indeed, technological backwardness are contributing causes for the emergence of oligarchical rule. The very existence of the young United States as a Federal Republic is a demonstration of this point. The average American was culturally and economically superior to the average Briton of the Eighteenth Century. Moreover, since nations which did not compete technologically would be strategically inferior, even the states committed to oligarchism, such as Britain, were compelled to adopt from technologically superior France that very same technological progress which they hated to see in French hands."[6]

However, by the turn of last century, there were large populations, especially in the technologically robust USA, that could no longer be subverted by an appeal to religious mysticism. As Carl Jung pointed out, a new pseudoscientific basis was needed for cult creation.

Enter H.G. Wells. Beginning in 1894, Wells wrote a dozen science fiction novels and over seventy short stories for the *Pall Mall Gazette*, owned by the Venetian Black Nobility family, the Astors. This is the same family that years later would sponsor Hitler along with other leading American and British oligarchs. The most important by far was his *War of the Worlds*, the still-definitive treatment of an invasion of earth by technologically superior beings, in this case from Mars. The point of the story is simple: man is threatened with extermination by an invading, superior force. The future is a Dantesque spiral leading us to Hell in a hand basket. Man finally defeats Martians, but not through our creativity and human intelligence, but rather by the natural action of bacteria, itself a lowest of low forms of life.

The sponsors of science fiction let their intentions be known: Men were beasts, technology was impotent, and the highest product of the centuries-long struggle of man to govern his affairs according to the law of progress, the sovereign nation state, was to be replaced by a universal feudalist order. Thus, "the essential method of 'dumbing down' of subject population is the vicious suppression of the creative powers of reason, those distinctively human mental capabilities which are expressed typically in the form of valid axiomatic-revolutionary discoveries in physical science."[7] With so much at stake, however, more than Wells was needed to transform a vibrant American population with visions of grandeur into a corral filled with docile collections of tamed human cattle. The strategy was simple and effective – retard creativity in all fields and slow down the rates of economic and scientific progress. Thus, it was in the editorial columns of science fiction magazines that a first critical theory of science fiction was put to the test.

Another character heavily involved in molding the minds of the young towards a specific goal outlined by the oligarchy was an emigrant from Luxembourg, Hugo Gernsback, a fellow of the British Royal Society who emigrated to the United States in 1904. In 1926, Gernsback founded *Amazing Stories*, itself the first specialist science fiction magazine in America, thus creating the first crop of America's science fiction writers.

Through magazines such as *Amazing Stories* and later *Astounding Stories*, science fiction "would boldly express what science did not yet dare to affirm or suggest, and even what it would find absurd. It would develop the social, ethical, metaphysical or purely logical consequences of science, or set out to confirm its ideological assumptions. In such a perspective, the literature of science fiction would be a continuation, albeit purely verbal and conceptual, imaginary and anticipative, of the real scientific work that itself sometimes uses fictions in the form of so-called 'thought experiments.'"[8]

The next big hit in the science fiction publishing belonged to a *Astounding Stories*, edited by John W. Campbell. Campbell was a physics graduate of MIT and Duke University and a close associate of U.S. Naval Intelligence operative L. Ron Hubbard, a small-time science fiction writer, would-be occultist and the founder of the cult of Scientology. Virtually every successful U.S. science fiction writer was groomed by Campbell, including Isaac Asimov, Theodore Sturgeon, A.E. Van Vogt, a leading Scientologist, Frank Herbert as well as British author Arthur C. Clarke in part due to the forcefulness of Campbell's personality by which he would bend his authors to his own objectives and in part because he paid more cents per word than anyone else in the business.

Campbell's entire stable consisted of authors with a strongly anti-scientific bias of science fiction and its links to the British establishment. For example, one of them was Robert A. Heinlein, an Annapolis graduate and a "skilled psychological warfare operative with close ties to Naval Intelligence that along with Air Force Intelligence has very close interfaces with the British intelligence services."[9]

In 1960, Heinlein was re-deployed into the MK-ULTRA drug proliferation project and in 1961, he came out with *Stranger in a Strange Land*. The theme deals with a Martian boy, Valentine Michael Smith, the son of astronauts, orphaned on Mars after the entire crew died. "Smith was raised in the culture of the Martian natives, beings with full control over their minds and bodies. A second expedition to the planet some twenty years later brings Smith 'home' to Earth. Since he is heir to the fortunes of the entire exploration party, which includes several valuable inventions, Smith becomes a political pawn in government struggles. On Earth, he withdraws into a trance-like state of another plane of existence,"[10] in other words, into a state of higher consciousness, whenever he is bothered by society. The book became the bible for the drug culture that was being organized by Gregory Bateson, Ken Kesey and Timothy Leary.

The most famous of the genre's writers is Isaac Asimov. A firm believer in Malthus' doctrine of depleting natural resources, a theme he addresses repeatedly in his works such as *The Caves of Steel*, for example, Asimov was brought in to work with Heinlein at the Naval Experimental Lab during World War II. Asimov's most important work is *The Foundation* trilogy, about two small elite groups, one specialised in high technology and the other in psychological warfare, "who tuck themselves away on a hidden planet and wait for the expected collapse of the galactic empire. The two groups are modelled on Britain's Tavistock Institute and the hidden Aldermaston labs, from which all scientific research and development on the British Isles is controlled."[11]

One of the few non-American authors in the Campbell stable was the legendary Arthur C. Clarke, a graduate of King's College at Cambridge and a veteran of the Royal Air Force radar program.[12] He was a conscious disciple of H.G. Wells and science fiction mystic Olaf Stapleton. Clarke's best known work is *2001, A Space Odyssey*. The novel's theme of transcendent evolution is attributed to the influence of British author Olaf Stapledon. It describes how all human progress is due to intervention from a superior alien race. Clarke is also the author of the best selling novel *Childhood's End*, in which a superior race of aliens called the "overlords," invades the Earth, ending all war, helping to form a world government, and turning the planet into a near-utopia from the safety of their space ships. Decades later, the "overlords" show themselves, judging humanity and deciding to merge the human race into the "disembodied cosmic vortex."

CARL SAGAN AND COSMOS

No matter how you look at it, there are only two sides to the UFO debate: you either believe it or you don't. About the only person the mainstream Establishment media trusted to speak of UFOs, astrology, and cults was the late Carl Sagan. In 1980, the landmark series *Cosmos* premiered on public television. Since then, according to the promotional material on the series, "it is estimated that more than a billion people around the planet have seen it. *Cosmos* chronicles the evolution of the planet and efforts to find our place in the universe. Each of the 13 episodes focuses on a specific aspect of the nature of life, consciousness, the universe and time. Topics include the origin of life on Earth (and perhaps elsewhere), the nature of consciousness, and the birth and death of stars. When it first aired, the series catapulted creator and host Carl Sagan to the status of pop culture icon and opened countless minds to the power of science and the possibility of life on other worlds."[13]

It doesn't take 13 episodes to establish, however, that pure science is not exactly what the series *Cosmos* is promoting. From jumping dolphins choreographed to Wagner's Valkyries to psychedelic galaxies and futurologist drawings, from singing whales and Hindu gods, there is a dialectical change from real science to Sagan's *Cosmos*, which takes us deep into the world of mysticism. In place of real science, which has always been an expression of mankind's moral commitment for human progress through mastery of the natural universe – *Cosmos* proposes an Aquarian-Dionysian version of science based on existentialist touchy feely environmental irrationalism where all involved would participate in a common cosmic consciousness. Where did Sagan's people draw the line between a psychedelic experience and quackery? In this sense, *Cosmos* copies the ideas from Friedrich Georg Junger's *The Perfection of Technology*, written in 1939. Junger writes: "Technology fills the air with smoke, pollutes the water, destroys forests and animals. This leads to a condition in which Nature must be protected from rational thought." In other words, what these kooks can´t understand about nature, must therefore have no explanation, except of course if you accept extraterrestrial ramblings as plausible scientific hypothesis.

The essence of science are the methods by which discoveries are made, called the method of hypothesis which was developed in ancient Greece. "The core of this method is the belief that the laws of the development of the universe are coherent with the human mind's own notion of lawfulness, so that the truth of the universe is fundamentally knowable to man."[14] Instead, Sagan adopts a consistent form of distortion in order to focus the audience's attention away from the coherence of man's mind. How does he do that? By attempting to break the universe down into a collection of objects, just as Euclid and Aristotle did before him. That's because ideas, to Carl Sagan, occur in a space devoid of reality. "Only thus, in unopposed emptiness could better elements be joined together in a continuous construction."[15] But, people are not nominalists on this account. Through the coherence of our mind, we know what we have experienced, and can identify existences as corresponding to such experiences without needing to break the experience into a collection of names because our real knowledge is that of our human, creative relationship to processes, not things.

For all man's science, proclaims Sagan, he is but a speck of dust in the vastness of the cosmos. Suddenly, we see "Sagan sailing through space in a flying saucer, assuring us that somewhere out in the vastness there exists life, perhaps superior civilization, trying to communicate with us, visit us...or perhaps they already have."[16] And just to make sure, we lower humans know our place in the universe, on page 313 of *Cosmos*, Sagan offers us a computer summary of extraterrestrial civilizations, and our place in 'the great chain of being.'

The barrier is breached between this world and the next, an initiatic rending of the veil of the unknown: space being seen as the domain of both the dead and the higher spiritual force. Suddenly, the vivid colors of the artist's drawing begin twisting, growing, shrinking, and changing colors, like a hallucination of an LSD acid trip.

In case the LSD and extraterrestrial do not work, try some Hinduism for a change. Again, let us refer to Cosmos' production for further information: "The Hindu religion is the only one of the worlds great religions dedicated to the idea that the cosmos itself undergoes an immense, indeed an infinite, number of deaths and rebirths... There is a deep and appealing notion that the universe is but a dream of a god, who after a hundred Brahma years (each 8.64 billions years long) dissolved himself in a dreamless sleep. The universe dissolves with him – until after another Brahma century, he stirs, recomposes himself, and begins again to dream the great cosmic dream. Meanwhile, elsewhere, there are an infinite number of universes, each with its own god, dreaming the cosmic dream."[17]

But Sagan doesn´t end there.

"These profound and lovely images are, I like to imagine, a kind of premonition of modern astronomical ideas ... It is by no means clear that the cosmos will continue to expand forever...If there is more matter than we can see – hidden away in black holes... then the universe will hold together gravitationally and partake of a very Indian succession of cycles, expansion followed by contraction, universe upon universe, cosmos without end."[18]

Instead of asking relevant questions that deal with mankind's place in the universe, Sagan takes us on a drug-induced space ride into oblivion. Through the power of television alchemy, *Cosmos* turned Sagan into a prophet. It is time that someone unmasked the idiocy of what fools had feared as the Mighty Wizard of Oz.

It is not surprising that this media extravaganza was funded by individuals and organizations closely linked to the same Aquarian Conspiracy crazies, like the Aspen Institute's Robert O. Anderson, who created the environmentalist movement through institutions such as Club of Rome, Friends of the Earth, WWF and World Watch Institute.

World Wildlife Fund, an organization created by the British Royal Consort, Prince Philip, aggressively portrays wild animals as 'better human beings,' compared to humans and modern industrial society. Amongst the people associated with WWF is the internationally renowned director of the Frankfurt Zoo, a man who once said that Earth would be lovable, if the human population were no more than 20 million.

But the excitement, the interest and the attention to real progress in space exploration, cannot be killed through kookery, science fiction, budget cuts or cultural warfare. What the mind benders at Tavistock don't seem to understand is that the human spirit is far more resilient than their zero growth, environmentalist agendas. We choose to explore space, not because it is easy, but because it is hard. This has always been man's destiny.

The *Cosmos* series was funded through a grant to the Public Broadcasting System of more than $1 million from the Atlantic Richfield Company. Chairman Robert O. Anderson and President Thornton Bradshaw are among the most prominent and influential American opponents of scientific, technological and cultural progress. Since the mid 1960s, the policy faction of which these two are members have been committed to the success of what was known in the 1960s as the Aquarian conspiracy, the undermining of society through the antiwar, environmentalist and drug and rock counterculture movements and is today committed to de-industrializing America to bring us back to the new Dark Age. The age of scientific rationale – characterized in Marilyn Ferguson's book *The Aquarian Conspiracy* as the Age of Pisces – will now be cut off by the Age of Aquarius.

For most of PBS' history, funding was provided by AT & T, a CFR company; Archer Daniels Midland, a Trilateral Commission company; PepsiCo, a CFR company whose Chief Executive Officer, Indra Krishnamurthy Nooyi is a Bilderberg and a Trilateral Commission Executive Committee member; and Smith Barney, a CFR company and one of the world's leading financial institutions. Therefore, you would hardly call PBS a non-profit, family oriented enterprise owned and operated by America's public television stations, would you?

What's more, Atlantic Richfield Company (ARCO), Anderson and Bradshaw have been particularly active in the environmentalist efforts of the Aquarian Conspiracy. Anderson personally funded the first Earth Day in 1970 to the tune of $200,000 to create the "grassroots" friends of the earth movement in opposition to technological progress. Hundreds of millions of dollars went into Earth Day 1970, a vast public relations stunt to get the green movement off the ground. Earth Day was bankrolled by the United Nations, Atlantic Richfield, the Ford and Rockefeller foundations and directed by the Colorado-based Aspen Institute of Humanistic Studies, one of the most important planning centers playing a dubious role against technological progress and the destruction of nuclear energy and spawning of the environmentalist zero growth operations.

Both Anderson and Bradshaw are leaders of Aspen Institute for Humanistic Studies. Aspen created the first antinuclear movement and devised the Malthusian environmentalist belief system that science and technology are intrinsically evil inflicted by man on nature. The two ARCO executives are also members of the exclusive Club of Rome, the leading international institution pushing polices of retrogression in technology, and reduction of the world's population by several billions of people – that's genocide, in case you didn't know it.

Founded in April 1968 by Alexander King and Aurelio Peccei, as a Malthusian organization with an agenda for genocide, the Club of Rome, established at a meeting of 30 individuals from ten countries, consists of the oldest members of Venetian Black Nobility of Europe, descendents of the richest and most ancient of all European families who controlled and ran Genoa and Venice in the XII century. In 1972, they published their manifesto on the Predicament of Mankind under the title, *Limits to Growth*, which showed that Earth could not sustain the current population expansion in the age of depleting natural resources. The final objective is the collapse of the world economy, even with their version of 'unlimited' resources, which includes zero technological advancement in science, or development of revolutionary new technologies. If you break through the Babylonian verbal confusion, their 1992 report, *The Global Revolution: A report by the Club of Rome Council* leaves little room for doubt as to their real agenda: "In search for a new enemy to unite us, we came up with the idea that pollution, the threat of global warming, water shortages, famine and the like would fit the bill." They concluded with the following, "The real enemy, then, is humanity itself."

In other words, the financial capitalists who launched a Carl Sagan from a college professor with bell bottom jeans to an intergalactic superstar are the organizers of the same antiscientific policy Sagan's series *Cosmos* peddles. Sagan's services were popularized as part of a willful attempt to eradicate scientific progress because the idea of ceaseless progress in favor of a better future is anathema to the oligarchs behind the curtain.

But the weirdness doesn't end there.

Sagan was one of the founding members of something called the Planetary Society. According to their own web page, "The Planetary Society, founded in 1980 by Carl Sagan, Bruce Murray, and Louis Friedman, inspires and involves the world's public in space exploration through advocacy, projects, and education. Today, The Planetary Society is the largest and most influential public space organization group on Earth. Dedicated to exploring the solar system and seeking life beyond Earth, The Planetary Society is non-governmental and nonprofit and is funded by the support of its members."

Sounds wonderfully promising, except that from the very beginning, the board of advisers included a group of the leading spokesmen for the Aquarian Conspiracy's environmentalist and de-industrial campaigns in the United States. Among them Isaac Asimov, the Malthusian science fiction writer; Norman Cousins, editor of the *Saturday Review* and member of the Club of Rome; John Gardner, chief of the environmentalist Common Cause organization and president of Rockefeller interfaced Carnegie Corporation; Shirely Hufstedler who sat on the board of Aspen Institute; and Lewis Tomas, leading advocate of euthanasia.[19] What's the common theme to their participation? Zero growth, zero progress idealism of its leading members.

THE EXTRATERRESTRIAL IMPERATIVE

Thirty years ago, space scientist Krafft Ehricke proved that the alternative of a no-growth policy would lead to a collapse in civilization: "geopolitical power politics, wars over natural resources, waves of epidemics, death-oriented population stabilization and extreme poverty." To combat the no-growth policy, Ehricke developed the concept of extraterrestrial imperative to confront the attack on mankind from the proponents of Malthusian theories of scarcity. The central concept Ehricke developed is "the distinction between multiplication and growth." In a 1982 article, Ehricke stated that "growth, in contrast to multiplication, is the increase in knowledge, in wisdom, in the capacity to grow in new ways." It literally turned Club of Rome's genocidal *Limits to Growth* report on its head. Ehricke pointed out that the Malthusians never considered "space exploration as domain of man's activity because to them, Earth is a closed system."[20]

That's because as Ehricke himself wrote in an 1973 article on his mentor Hermann Oberth, "The development of an idea of space travel was always the most logical and the most noble consequence of the Renaissance ideal, which again placed man in an organic and active relationship with this surrounding universe and which perceived in the synthesis of knowledge and capabilities its highest ideas... The concept of 'limit' and 'impossibility' were each relegated to two clearly distinct regions, namely the 'limit' of our present state of knowledge and the 'impossibility' of a process running counter to the well understood laws of nature."[21]

Countering the anti-science description of the view of the Earth by the Apollo astronauts on their way to the Moon, as a 'fragile' globe that mankind is 'destroying', Ehricke wrote, "Earth is not merely a spaceship. It is a member of the Sun's convoy traversing the vast ocean of our Milky Way galaxy... Our companion worlds are underdeveloped. Earth is the only luxury passenger liner in a convoy of freighters loaded with resources. These resources are for us to use, after Earth has hatched us to the point where we have the intelligence and the means to gain partial independence from our planet."[22]

Indeed, there are whole new worlds for us to explore, about which we know very little; an exploration, which should keep us busy for a few hundred generations. And this is exactly what the anti science platform through Sagan and Aspen and WWF and H.G. Wells has been trying to avoid at any cost.

THE COMING OF THE UFOS

We are back to where we started, staring at the 'doors of perception', enmeshed in a labyrinthine task of connecting the dots. From past chapters, we connected the dots to the involvement of intelligence agents in the field of Egyptian archeology to the study of the paranormal, OSS operations, psychological warfare, Paperclip scientists, JFK assassination... a kind of grab bag of government-approved operations and would include everything from séances and witchcraft to remote viewing and exotic drugs. And either we agree with JFK prosecutor Jim Garrison that such coincidences are evidence of an intelligence operation or that the coincidences represent something else: an "acausal connecting principle" to use the phrase invented by Carl Jung, that nonetheless demonstrates important linkages between events in the world.

For example, Hungarian-American physicist, Theodore von Kármán and NASA team first test rocket in the Arroyo Seco in Pasadena on October 31, 1936. October 31, the night of Halloween and coincidentally, the same day as the last Houdini séance. Those who think that perhaps the first rocket launch on Halloween was simply a coincidence should refer to the official JPL website where they state that the American space age began on "January 31, 1958 with the launch of the first US satellite, Explorer I, built and controlled by JPL." And as Peter Levenda aptly connects the dots in his underground best seller, *Sinister Forces*, "January 31, of course, is the pagan holiday known to Christianity as Candlemas and to pagans as Oimelc. Thus, for better or for worse, meaningful or not, the launch that took place on Halloween, 1936 culminated in the launch on Candlemas, 1958, and was the result of the work of the same organization, for von Kármán's GALCIT team eventually became the founding members of the Jet Propulsion Laboratory. The man in charge of the Explorer launch was Dr. Wernher von Braun, the Nazi scientist formerly of Peenemunde and Nordhausen."[23]

1947 is a pivotal year in the postwar period. This was the year CIA was created; the year the Dead Sea Scrolls were discovered and the year Winston Churchill made his famous "Iron Curtain" speech, thus declaring the beginning of the Cold War. It is the year that US Navy begins project CHATTER, the search for a viable truth serum, a magic potion to unlock the secrets of the mind. It is the year that the greatest Satanist of the 20th century, Aleister Crowley died. And, if we are to believe the Beatles, it is the year that Sergeant Pepper's Band learned to play.

It was the year of the famous Roswell, New Mexico UFO crash. More – or less – than meets the eye? Extraterrestrials, spaceships, 'men in black,' visitors from others planets and the UFO phenomenon disembarked onto the front pages of Americans newspapers and in no time all Hell broke loose.

The 1947 UFO Roswell sighting attracted two men – a former OSS officer Fred Lee Crisman,[24] and former FBI Special Agent Guy Bannister – who both would come under suspicion twenty years later for their supporting roles in the Kennedy assassination. "Crisman and Bannister, two men friendly with a large list of people who later would be a part of the conspiracy to assassinate the President. The odds against this happening must be astronomical. It is the constant appearance of 'coincidences' like these that leave most people apoplectic, speechless with disbelief and gazing on the world around them with haunted, suspicious expressions, as if reality itself were layered like an onion, a palimpsest on which numerous events were written over each other, all on the same page."[25] In this case, we have Operation Paperclip, UFOs, and the Kennedy assassination all written on the same sheet of onionskin parchment. Nazis, aliens and political murder. One can almost sympathize with Pontius Pilate, who asked "What is the truth?" – and the temptation to wash one's hands of the whole matter is almost too strong.

One of the accusations levelled at UFO eyewitnesses is that their testimony is tainted by reason of inexperience, lack of professionalism, lack of knowledge of astronomical phenomena, etc. In other words, they are not scientists, military, police or establishment with credentials, that is the people in charge of determining the contours of our everyday reality. What would happen if you could parade a perfect witness or a series of them in front of the worldwide audience? Would that make a difference in how we perceive the UFO phenomenon?

CREATING CONSENSUS

The role of Air Force Intelligence is revealed by examining the pattern of recorded major UFO sightings worldwide between 1947 and 1973. "Of the 145 U.S. sightings, fifty four were claimed by the military: forty six by the Air Force, six by the Navy, one by the Army, and one by the Coast Guard. The modus operandi is clear. A couple of Air Force colonels announce they have seen a flying saucer. This makes headlines, and suggestible people everywhere begin interpreting every flying light they see as a UFO. Such UFOs make headlines in national papers which leads to more and more stories on UFOs. In case things slow down, a top Air Force official steps forward and announces that he is resigning from the force and will reveal all the information about alien spacecraft that the military has secretly kept hidden. The resignation of Vice Admiral Roscoe Henry Hillenkoetter, United States Navy, a decorated war hero and the former director of the CIA, in 1960 is one such case." Hillenkoetter quit his job at NICAP (National Investigative Committee of Aerial Phenomena, the most prestigious of all UFO investigative groups) within days of a CIA investigation of his proposed role in opening the UFO investigation at the Congressional level through the Senate Science and Astronautics Committee with support of senators like Barry Goldwater. "Behind the scenes, high-ranking Air Force officers are soberly concerned about the UFO's," Hillenkoetter said.[26]

UFO promoter Jacques Vallee describes the effects of the UFO hoax in detail in the pages of his Messengers of Deception:

"1. The belief in UFOs widens the gap between the public and scientific institutions.

"2. The contactee propaganda undermines the image of human beings as masters of their own destiny.

"3. Increased attention given to UFO activity promotes the concept of political unification of the planet.

"4. Contactee organizations may become the basis of a new 'high demand' religion.

"5. Irrational motivations based on faith are spreading hand in hand with the belief in extraterrestrial intervention.

"6. Contactee philosophies often include beliefs in higher races and in totalitarian systems that would eliminate democracy... Strong belief in extraterrestrial intervention could revive this primitive concept, with particular groups claiming privileges peculiar to those who descend from stellar explorers."[27]

Vallee is a close working associate of Profesor Ron Bracewell, who along with MK-ULTRA operative John Lily, astro-mystic Carl Sagan and NY Times science editor Walter Sullivan, author of We Are Not Alone, is a member of the Order of the Dolphin, founded in 1961 to promote the idea of the existence of superior or, extraterrestrial civilisations. Another kook who belongs to this group is American astrophysicist, Frank Drake, founder of SETI, a group whose primary objective is the Search for Extra-Terrestrial Intelligence by using the Drake equation, an equation used to estimate the number of detectable extraterrestrial civilisations in the Milky Way galaxy.

THE END

There is something enigmatic about asking questions. Questions have a way of framing the future. Through *The War of the Worlds,* H.G. Wells asked the following: "Well then, someone may say, if there are globes in the heaven similar to our earth, do we vie with them over who occupies the better portion of the universe? For if their globes are nobler, we are not the noblest of rational creatures. Then how can all things be for man's sake? How can we be the masters of God's handiwork?"

I have to admit these wouldn't have been my questions. I would want to know, "How do we change the world?" "How do we become immortal?"

In the late 1950s, scientist Lloyd Berkner, who had accompanied Admiral Richard Byrd on his Antarctic expedition in 1928, stated his vision for the future: "Science is creative beauty in the highest sense. It provides a systematic and reliable criterion of universal applicability in Plato's search for the 'harmonious, the beautiful, and the desirable....' Truly, the characteristic of civilized man that distinguishes him from all other creatures is his learning, the ability to utilize knowledge to free himself from the vicissitudes of his environment."

Long before Club of Rome became a nightmare reality, Berkner anticipated their argument regarding the 'cost-benefit' of space exploration: "Each new technology derived from science has a permanence that continues to benefit society indefinitely in the future. Thus capital represented by discovery outlives all other forms. Consequently, the investment in basic research should be written off over an indefinitely long time against the permanent gains acquired by society."[28]

This is not what the oligarchy and Tavistock wanted to hear. In the mid-1960s, the Tavistock Institute's magazine, *Human Relations* reported, with alarm, that the space program was producing an extraordinary number of 'redundant' and 'supernumerary' scientists and engineers. "There would soon be two scientists for every man, woman, and dog in the society," they warned. "The expanding pool of these scientists and engineers would have a profound impact on the values of American society, from skilled workers to office clerks,"[29] Tavistock reported.

Between the late 1950s and late 1960s, people all over the world, watched in awe, as man conquered space. And as Marsha Freeman explains in her great expose on *Mankind's Next 50 Years Of Space Exploration*, "It was not only scientists and engineers, and children who, by the thousands joined rocket clubs, spending weekends launching amateur rockets, who were infected with the optimism of the Apollo program. In 1962, the editors of *Fortune* magazine, reflecting the view of American industry, published a book, *The Space Industry: America's Newest Giant*, which included a chapter titled, "Hitching the Economy to the Infinite":

"There is no end to space, and so far as the U.S. economy is concerned, there will probably be no end to the space program.... Man has hitched his wagon to the infinite, and he is unlikely ever to unhitch it again.... The space venture, in short, is likely to be more durably stupendous than even its most passionate advocates think it will be.... Overall, nothing is more fecund, industrially and socially, than large mobilisations of scientific knowledge and effort; and this is the greatest mobilisation of them all."[30]

Our mortal life has a beginning, and an inevitable end. What, then, is our immortal *interest* in being a person? Can we consider immortality to be an approximation of infinity? "When we generate, and transmit, discoveries of principle, to our children, to those who come later, we live forever in the history of mankind. Our mortal existence is no longer a matter of a beginning and an end: Our mortal existence is a place in eternity, from which we radiate the experience of generations before us, and radiate our existence into the future."[31]

In a speech in Texas, on Jan. 19, 2007, NASA Administrator Mike Griffin gave his view on space exploration. The real reason we strive to better ourselves is because "we want to leave something behind for the next generation, or the generation after that, to show them that we were here, to show them what we did with our time here. This is the impulse behind cathedrals and pyramids and many, many other things.... It is my observation that when we do things for real reasons, we produce our highest achievements."

Griffin continued: "The cathedral builders knew that reason... We look back across 600 or 800 years of time, and we are still awed by what they did...."

He ended by saying, "It is my contention that the products of our space program are today's cathedrals. The space program addresses the real reasons why humans do things. The cathedrals of the second 50 years of the Space Age are waiting to be built. This will require nothing less than the philosophical view of mankind which created the cathedrals of centuries past, and of the first half-century of space exploration."

Against the generalized euphoria of legitimate space exploration and the betterment of mankind against future generations, the Aquarian-Dionysian version of science based on existentialist environmental irrationalism, psychedelic experience, quackery and common cosmic consciousness stands exposed for what it is – a sophisticated tool designed, on the one hand, to entrap, disorient, and destroy the creative minds of the younger generation who have 'dangerous' illusions that are associated with belief in the idea of technological progress and the spirit of innovation and on the other, to displace man from his rightful place in the Universe – the epicenter.

Truth always lies in the higher order of processes. True sovereignty lies not in popular opinion, but in the creative powers of the individual human mind. The only defense we have against the mind benders is to develop our own mental capabilities – individually – as a firewall against psychological warfare and cultural colonialism. Demons, God... Aliens and us.

And so it goes.

Endnotes

1. Hugo Gernsback, Editorial, *Science Fiction Weekly*, 1930, In Gary Westfahl, *Hugo Gernsback and the Century of Science Fiction* (2007), 166

2. Robert Zubrin, There is No Science in Science Fiction, *The Campaigner*, April 1981, p.25

3. The `No-Soul' Gang Behind Reverend Moon's Gnostic Sex Cult, Larry Hecht, *EIR*, December 20, 2002

4. Carol White, *The New Dark Age Conspiracy*, New Benajim Franklin House, 1980 p.5

5. Ian D. Colvin, *The Unseen Hand in English History,* Nabu Press, 2010, reproduction of a book published before 1923

6. LaRouche, *How Bertrand Russell Became an Evil Man*, Schiller Institute, July 28, 1994.

7. LaRouche, *How Bertrand Russell Became an Evil Man*, Schiller Institute, July 28, 1994

8. Gerard Klein, *Learning From Other Worlds, Estrangement, Cognition and the Politics Of Science Fiction And Utopia*, Liverpoool Universwity Press, 2000, p.11

9. Robert Zubrin, There is No Science in Science Fiction, *The Campaigner*, April 1981, p.28

10. http://en.wikipedia.org/wiki/Stranger_in_a_Strange_Land

11. Robert Zubrin, There is No Science in Science Fiction, *The Campaigner*, April 1981, p.28

12. http://en.wikipedia.org/wiki/Arthur_C._Clarke

13. Carl Sagan, Cosmos, KCET productions, 1978-1979

14. Robert Zubrin, Carl Sagan's Kook Cosmos, *The Campaigner*, February 1981, p.8

15. Helga Zepp-LaRouche, The Historical roots of green fascism, *EIR*, April 20, 2007, p.29-38

16. Robert Zubrin, Carl Sagan's Kook Cosmos, *The Campaigner*, February 1981, p.10

17. Carl Sagan, Cosmos, KCET productions, 1978-1979

18. Ibid

19. http://medicine.jrank.org/pages/602/Euthanasia-Senicide-modern-argument.html

20. Why Man must explore Mars, *EIR Science & Technology*, Volume 24, number 30, July 25, 1997

21. Ibid

22. Ibid

23. *Sinister Forces – The Manson Secret: A Grimoire of American Political Witchcraft*, Peter Levenda, Trineday Press, 2001

24. See FBI files: BUFILE 62-83894-52, 62-83894-40, -89, -91 all with the designation "SM-X" or "Security Matter – X" or "SM Dash X."

25. *Sinister Forces – The Manson Secret: A Grimoire of American Political Witchcraft*, Peter Levenda, Trineday Press, 2001

26. http://www.phils.com.au/nytarticle.htm

27. Robert Zubrin, There is No Science in Science Fiction, *The Campaigner*, April 1981, p.30

28. Marsha Freeman, Mankind's Next 50 years of Space Exploration, *EIR*, October 19, 2007

29. Ibid

30. Mankind's Next 50 Years Of Space Exploration, Marsha Freeman, *EIR*, October 19, 2007

31. How To Reconstruct A Bankrupt World, Lyndon LaRouche delivered the following address to a public event sponsored by Schiller Institute on Dec. 12, 2002 in Budapest, Hungary

Index

233

Josiah Macy, Jr. Foundation   91, 183,
    185-188, 191, 193-195, 199, 201,
    203, 208-210
Jung, Carl   42, 44, 130, 132-134, 137,
    142, 168-169, 177, 199-200, 205,
    216, 226
Junger, Friedrich Georg 220
Jureidini, Paul 46

K

Kármán, Theodore von 226
Karzai, Hamid 127
Keith, Jim 83, 85, 118
Kennan, George 35
Kennedy, John Fitzgerald 41, 50, 66-75,
    77, 82, 85, 100, 227
Kepler, Johannes 154, 205
Kesey, Ken 85, 99, 111, 218
KGB 17, 19-21
King, Alexander 223
Kissinger, Henry 100, 201
Kline, Nathan 14, 31, 204
Kornfeld, Artie 110
Kravis, Marie-Josée 56
Kristol, William 146

L

Lady Gaga 79, 124
Laing, R.D. 104, 120
Lanier, Jaron 206
Lansing, Robert 51
LaRouche, Lyndon 119-121, 129, 177-
    178, 208-209, 232
Last Days of Pompeii, The 93
Laszlo, Ervin 114
Lawrence, D.H. 97
Lazarsfeld, Paul 155, 191, 194
Leary, Timothy 86, 98, 112, 218
LeBon, Gustav 130, 131, 177
LeDoux, Joseph 188
Lehrer, Jim 146
Leibniz, Gottfried Wilhelm 189
Lennon, John 117-118, 120
Levenda, Peter 9, 22, 38-39, 41, 43-44,
    46, 58-59, 90, 109, 119, 121, 178,
    232

Lewin, Kurt 9-11, 23, 38, 42, 49-50, 104,
    140, 178, 185, 190-192, 195, 199
Life 86
Lily, John 228
Limits to Growth 223, 225
Lincoln, Abraham 70, 72, 179
Lindsay, Franklin A. 35
Lippmann, Walter 151-152
Lockwood, Joseph 110
Lord, Winston 147
Lovecraft, H.P. 81
Luce, Henry 86
Lucifer 91, 109

M

Mackey, Albert 74, 83
Mackinder, Halford 214
Maeterlinck, Maurice 22-23, 39
Malraux, André 165
Malthus, Thomas 187, 199, 210, 218
Mankind's Next 50 Years Of Space Ex-
    ploration 230, 232
Manson, Charles 38, 45, 76, 108, 109,
    121, 232
Marcuse, Herbert 87
Marcus, L. 8, 27, 38, 102, 113, 120
Marshall, George 34
Mass Psychology 91, 131, 134
Mathews, Jessica Tuchman 56
Maxwell, James 191
McCulloch, Warren 185-187
McGlothlin, W.H. 86
McGregor, Douglas 184
McMurtry, John 3, 5
McNamara, Robert 197
Mead, Margaret 99, 114, 122, 183, 186,
    194-195, 199-200
Mein Kampf 131
Mengele, Josef 49, 196
Metropolis 211
MI5 17-21
MI6 17-21, 45, 118, 120, 122
Milner, Alfred 196, 214, 215
Minnicino, M. 27, 29, 38, 40, 85-86, 99,
    118, 120, 154, 179, 208-210
Minor, Newton D. 145